Ana Carolina Feitosa de Vasconcelos
Lúcia Helena G. Chaves

Industrial waste from granite sawmills in soil fertility

Ana Carolina Feitosa de Vasconcelos
Lúcia Helena G. Chaves

Industrial waste from granite sawmills in soil fertility

An agronomic and environmental application

ScienciaScripts

Cover image: www.ingimage.com

This book is a translation from the original published under ISBN 978-3-330-75457-7.

Publisher:
Sciencia Scripts
is a trademark of
Dodo Books Indian Ocean Ltd. and OmniScriptum S.R.L publishing group

120 High Road, East Finchley, London, N2 9ED, United Kingdom
Str. Armeneasca 28/1, office 1, Chisinau MD-2012, Republic of Moldova, Europe
Printed at: see last page
ISBN: 978-620-8-25510-7

CONTENTS

SUMMARY

Brazil has large reserves of granite and a great potential for its industrialization. However, the processing industries face problems with the generation of large quantities of waste, which has been causing environmental impacts. In order to find alternatives for the use of these residues, the aim of this study was to evaluate the effects of these residues on the chemical properties of acidic soils and as a source of nutrients for plants. To this end, two experiments were set up in a greenhouse and in the Irrigation and Salinity Laboratory (LIS) of the Department of Agricultural Engineering at the Federal University of Campina Grande, in two different stages, using soil from the Brejo Paraibano region, which was removed from the arable layer (0-20 cm), homogenized and chemically analysed. Samples of the granite sawdust sludge were collected from the industry's deposit tank and subjected to chemical and mineralogical analysis. The doses of granite sawdust slurry used in the first stage were determined on the basis of the amount of calcium carbonate needed to raise base saturation to 60%; in the second stage, the doses were based on the calculation for raising soil base saturation to 90%. The experimental units consisted of soil samples mixed and homogenized with their respective treatments and subjected to an incubation period with humidity corresponding to 80% of field capacity. The experimental design was entirely randomized with three replications. At the end of the incubation period, soil samples were collected from the experimental units of each experiment for chemical analysis. After this period, in the second experiment, corn (*Zea mays* L.) was introduced as a test plant. Each experimental unit consisted of two plants which, sixty days after planting, were subjected to determinations of leaf area, leaf length and dry and fresh biomass. The dry biomass was analyzed for nitrogen, phosphorus and potassium content. The chemical characterizations of the experimental units were carried out at the end of each experiment. Although the pH of the soil was not raised to adequate levels, the granite sawdust slurry had an acid-neutralizing effect and increased the CEC, phosphorus and organic matter of the soil. The granite sawdust sludge contributed to an increase in potassium levels in the plants. Granite sawdust sludge can be used for agricultural purposes.

KEY WORDS: industrial waste, corrective, fertilizer.

1-INTRODUCTION

Since the dawn of history, waste products and various residues have formed as a result of anthropogenic action, and when they are washed into rivers or into the air, they prove to be toxic, or at least a nuisance, compromising the environment. It was initially believed that the rapid increase in waste from industrialization and food production could be eliminated by distributing the polluted water and gases over as large an area as possible, in order to achieve optimum dilution of the toxic substances.

The time needed for decontamination or for pollutants to deteriorate spontaneously was wrongly estimated. Today we recognize the problem posed by environmental damage. A complete detoxification of all industrial emissions or discharges is so expensive that it is impossible to apply it across the board. It is necessary to reuse these industrial wastes in order to reduce the impact on the environment, since reducing production in some branches of industry to avoid greater pollution would have negative economic consequences (Fellenberg, 1980).

For use in agriculture, some industrial waste has already been studied in an attempt to reduce the negative effects of modernization, such as: urban waste, bauxite residue, boiler ash, basalt dust, sludge from the pulp and paper industry, bentonite (bofe) and others.

Brazil has large reserves of marble and granite with various aesthetically different types and great potential for the industrialization of these mineral resources. According to the Brazilian Mineral Yearbook (1998), granite reserves in the state of Paraiba total 144,874 m3.

In the northeast of the country, granite sawmills have been worrying owners, environmentalists and government officials due to the growing amount of waste in the form of the mud resulting from the action of the looms, which may be responsible for the direct contamination of rivers or the soil itself. When it flows, the mud drowns plants and animals, damages soils, poisons watercourses and springs, and when it dries, it is harmful to the health of humans and animals because it is breathed in as dust. The lagoons or settling ponds used by those industries whose administrations are more aware of environmental problems soon become full of sludge and have to be drained at great inconvenience and cost.

In order to develop and accelerate economic growth in this sector, without compromising the environment and the community in general, the exploitation of these rocks must be done rationally through proper management of the waste generated by this activity.

The Polimento de Granitos do Brasil S/A - POLIGRAN industry, located in Campina Grande - PB, produces 72,000 m^2 of flat granite slabs every year, with prospects for expansion. These are extracted from its own quarries, located in various municipalities in the state of Paraiba.

The granite sawing sludge (waste) generated in this industry, 66 m^3 /month, is basically composed of water, lime and grit and has been discarded and consequently accumulated in sedimentation ponds in the industrial area itself.

In order to work in the long term, with respect for the environment and human beings, several studies have been carried out on the reuse of this waste. Some of this material has been used for landfills in civil construction, as well as being applied as an alternative material in the manufacture of soil-cement bricks (Da Silva, 1998). Other studies are trying to make its use viable in the manufacture of refractory materials. Its chemical characteristics and high pH value indicate that it could be used in agriculture as a possible supplier of nutritional elements for plants or as a soil acidity corrector, which should be the subject of further research.

Based on this, this work was carried out with the general aim of studying the potential for agricultural use of the tailings from the industrial activity of sawing granite blocks, making it possible to reuse this material, thereby reducing the environmental impact generated by this activity and at the same time providing the granite processing industry with an alternative way of attributing value to the tailings. Specifically, to evaluate the effects of granite sawdust sludge as a corrective for acidic soils, thus improving the soil's chemical characteristics, and as a source of nutrients for plants.

2-LITERATURE REVIEW

2.1- The Man-Nature Relationship

2.1.1- A Brief History

In early modern times, human beings searched for their place in nature, trying to discover their role in the relationship with other species. Initially, his main idea was to exploit all animals and plant species for his own personal satisfaction, without worrying about ensuring their perpetuation.

With the passage of time, a new conception was created about this fact: God had planned the existence of all natural species and they had to exist and relate to each other in harmony. This idea sensitized a growing number of people and changed their behavior towards nature, giving rise to the idea of environmental preservation and that the exploitation of natural resources could not be done indiscriminately.

Agriculture and animal husbandry made great progress. Agricultural production was developed to meet the commercial interests of the time. Weeds, which had previously been eliminated to prevent them from compromising crops, were now researched for use in medicine. Animals, as well as serving as a means of subsistence, came to be seen as objects of ornamentation and prestige, depending on their rarity.

These processes were aspects of a much broader change in man's relationship with nature. The growth of cities and the intensification of the strong difference between urban and rural life meant that the city became a symbol of civility and the countryside of rudeness and rusticity.

However, in England, around 1800, the idea that the countryside was more beautiful and pleasant than the city was quite common among those who lived there. In part, this conviction was due to the deterioration of the urban environment. The coal burned in the early modern period contained twice as much sulphur as the product used today and its effects were proportionally lethal. Equally harmful was the pollution caused by the gases and waste generated by brewing beer, dyeing clothes, making gum and bricks, and all the other industries set up in the middle of cities (Thomas, 1988).

This pollution intensified as human action destroyed more and more of nature, extinguishing species according to the prevailing agricultural needs of the time.

2.1.2- Environmental Crisis

The increase in the world's population makes it necessary to create more sources of work, with natural resources being used at a greater rate, and waste also increasing.

The current crisis has given rise to a style of development that manifests itself in the processes of agricultural modernization, urbanization and changes in the resource base and energy use.

Environmental deterioration is an inevitable consequence of human progress, but it is a characteristic of the type of economic development that presents ecological unsustainability, inequality and social injustice. It is recognized that no social system can remain viable for long if its natural environment is damaged or destroyed, or if there is no protection against the deterioration of that environment.

An environmental policy aimed at conservation and environmental protection must take into account those who depend on natural resources for their survival. Any development policy aimed primarily at increasing the production of goods, without considering the sustainability of the resources on which production is based, leads to a decline in productivity, generating poverty.

Trade and industry play an important role in a country's economic and social development. The policies and operations carried out in these sectors form the necessary basis for reducing the impact on the use of resources and the environment through more efficient production processes, preventive strategies, cleaner production technologies and procedures to minimize or avoid the generation of waste.

The major ecosystems are dramatically threatened and the most affected areas of the planet are the underdeveloped ones. Without a prior solution to the serious socio-economic problems that ensure a reasonably decent life for the poor, little can be done to prevent them from also contributing to the deterioration of natural resources (Guimaraes, 1991).

For centuries, humanity had become accustomed to living on an unlimited plane. When the environment deteriorated to the point where it wasn't as sustainable as before, there was always a new frontier to expand exploration activities to. Today it is clear that natural reserves are limited in terms of resources or places to dispose of waste. Since the middle of the century, the world has lost a fifth of its tropical forests. Every year tons of humus are lost due to erosion, desertification, salinization and other soil deterioration processes, representing an area of agricultural use equivalent to the total territory of the Caribbean countries (Guimaraes, 1991).

The agricultural revolution laid the foundations for the emergence of cities and the Industrial Revolution created an artificial environment dependent on natural resources. Just as technology has improved the human environment in many ways, it has also led to drastic changes in the natural environment.

The increasing and exaggerated use of agricultural inputs, heavy and concentrated mechanization and the predominance of production aimed at foreign markets, which has stimulated the formation of vast areas of monocultures, have caused immense ecological and environmental problems.

In addition to the destruction of the soil and the indiscriminate use of pesticides, the clearing of forest reserves has given rise to activities such as extensive livestock farming, the exploitation of charcoal as a source of energy for steel mills and mining operations that contaminate river water with mercury and encourage the silting up of river basins.

In this context, it is of fundamental importance that the role of science and technology is present in human affairs for the implementation of policies and programs appropriate to sustainable development, and that the scientific and technological community is guaranteed the independence to investigate and publish its findings without restriction.

Multidisciplinary approaches and interdisciplinary studies among the scientific and technological community, decision-makers and the general public are the basis for raising awareness that economic development must be geared towards preserving natural resources and guaranteeing their existence for future generations.

The environmental crisis cannot be reduced to a question of keeping the air, water or soil clean, nor can the environment and development be conflated, since the former is the result of the latter. Environmental problems are the problems of unequal development for human societies and harmful development for natural systems.

2.1.3- Attempted solutions

In an attempt to find solutions to minimize the environmental impact caused by the disposal of urban and industrial waste, various segments of society have been engaged in developing research aimed at the possibility of using it economically. In this context, the possibility of the agronomic use of different types of waste has been investigated.

Fontes and Ribeiro (1999) cite studies carried out with steel slag associated with natural phosphates that proved the effect of this material as a soil acidity corrector. This slag treated with activated sludge also worked effectively for this purpose.

Phosphate sludge is a waste product from the steel industry which, because it contains around 30% P_2O_5, has been tested as a source of nutrients for plants in experiments with sorghum (Defelipo *et al.*, 1992, cited by FONTES and RIBEIRO, 1999).

The niobium processing industry produces a residue that corrects soil acidity by reducing the concentration of exchangeable aluminium in the soil and increasing pH, calcium and magnesium content and base saturation (Schunke *et al.*, 1998).

By-products from the zinc industry, such as ash and dregs, can be used as raw materials for fertilizers (Ponchio and Ballio, 1998, cited by FONTES and RIBEIRO, 1999). Amaral *et al.* (1996), evaluating the influence of these by-products on the chemical characteristics of the soil and on the growth and chemical composition of maize plants, found that this material acted efficiently as a corrective, raising the pH and neutralizing the exchangeable aluminium in the soil, as well as increasing its calcium and magnesium content.

There was also an increase in dry matter production and in the content of calcium, magnesium and zinc in the aerial part of the corn.

Residues from the pulp industry applied to eucalyptus plantations have increased soil fertility, as well as improving porosity and water retention capacity, increasing pH and soil P, K, Mg and Ca contents; they have enabled significant gains in wood volume when used with mineral fertilizer in forest plantations (Guerrini and Moro, 1994, cited by FONTES and RIBEIRO, 1999).

Urban waste that can be used as a source of nutrients for plants is the sludge generated by wastewater treatment plants (WWTPs) and organic waste compost, which generally has high levels of metals, especially zinc, lead and copper. Because it contains materials that decompose more slowly, compost has less fertilizing value than sludge from sewage treatment plants. The use of sewage sludge and compost in agriculture is a disposal practice that is becoming increasingly accepted (Tedesco *et al.*, 1999).

Basalt dust, which comes from the process of processing basalt rock for use in civil construction, has been used in agriculture in European countries and the USA. In Brazil, studies by Pereira and Kiehl (1999) show that its application to the soil increased dry matter production and the accumulated amounts of N, P, K, Ca, Mg, Cu, Mn and Zn in the plant. Novelino *et al.* (1995) obtained satisfactory results in reconstructing the fertility of soils in the cerrado region, especially in areas close to quarries exploited for the production of crushed stone.

Organic waste is generally produced by agro-industries, leather and footwear activities, oil extraction, among others. It has a high organic load and decomposes quickly when applied to the soil, helping to improve its fertility. The organic waste removed from decommissioned salt pans in the municipality of Cabo Frio- RJ improves the physical and fertility conditions of sandy soil and can be used as an organic fertilizer and soil conditioner (Araùjo and Perlingeiro. 1999).

Soil is the unconsolidated weathered surface of the earth's crust, containing organic matter and living beings. From a physical point of view, soil is a heterogeneous system, made up of solid, liquid and gaseous phases. From a compositional point of view, soil has an inorganic or mineral fraction, which generally predominates, and an organic fraction. Soils are formed in nature as a result of five factors: source material, climate, relief, time and organisms (Raij, 1991).

The source material is the most important factor in soil formation and its initial composition will determine its texture, mineral composition and nutrient richness. The ways in which rocks break down give rise to specific types of soil. In mechanical weathering, where water, temperature, wind and vegetation act, granular soils (pebbles and sand) are formed; on the other hand, when chemical weathering of rocks occurs, through oxidation, hydration and carbonation, fine soils are formed (Carvalho, 1989). Source materials rich in nutrients give rise to soils of high natural fertility, while those devoid of nutrients, such as sedimentary rocks impoverished by the intense weathering to which they have been subjected, give rise to poor soils.

In addition to the material of origin, other factors influence the volume and level of this natural reservoir of nutrients, which is the soil, classifying it from well supplied (fertile soil) to poorly supplied (poor soil). However, even fertile soils do not have an inexhaustible reserve of nutrients, and there is always a need to replenish what has been removed by the plants. The practice of fertilizing and liming has been an important factor in agricultural production as it adapts the soil to the ideal conditions for cultivation (Oliveira, 1991).

The cations released by weathering and the decomposition of organic matter differ in their charge, size and ability to polymerize. They respond differently to the ions and surfaces found in the soil. Some cations can precipitate as soil minerals; they can remain in the solution adsorbed to the surfaces of the solid phase (exchangeable cations) or they can remain in the soil solution (soluble cations). Aluminum generally precipitates as hydroxide or as aluminosilicates and is absent from the soil solution (Chaves, 1997).

Exchangeable cations are defined as those released from the surface of soil colloids by neutral salt solutions (extractant solutions). Because they are adsorbed, these cations are temporarily protected from leaching and, through cation exchange, are released into the soil solution, making them available to plants. Cation exchange is a reversible process. The cations can move freely, not occupying a fixed position, since the attraction between the cations and the surface of the colloids is simply through the charges.

Calcium is the predominant cation in almost all soils; sodium can accumulate in the soils of arid and semi-arid regions, but in most soils, it is the exchangeable cation that occurs in the smallest quantities. Exchangeable aluminum (Al^{3} +) can be present in considerable quantities in acidic soils (pH<5.5), as can exchangeable hydrogen ($H+$) (pH<4.0). The hydrogen corresponding to part of the soil's potential acidity is in a poorly dissociated form in the soil and does not take part in ion exchange reactions, so it is not considered an exchangeable cation.

One of the most important properties of soils for agriculture is their cation exchange capacity (CEC). This means that soils retain cations such as calcium, magnesium, potassium and others in smaller quantities, preventing them from following the drainage water, while keeping them available for plants. The CEC indirectly corresponds to the sum of the exchangeable bases (calcium, magnesium, potassium and sodium) plus hydrogen and aluminum. The part of the CEC occupied by these bases expresses the base saturation of the soil, which, together with exchangeable aluminum and exchangeable hydrogen, represents the operative part of the CEC in the soil under natural conditions (Raij, 1991).

Among the macronutrients, three are in the form of cations: potassium, calcium and magnesium; phosphorus and sulphur are in the form of orthophosphate and sulphate anions, respectively. Nitrogen is the only macronutrient that can occur in the soil in the form of a cation (ammonium) or another anion (nitrate). Hydrogen and aluminum stand out in terms of acidity and reaction in acidic soils.

Another important property of soils is their reaction, which can be acidic, neutral or alkaline. This reaction, assessed by determining the pH, cannot be underestimated since it influences the availability of all the chemical elements present in the soil, thus affecting the development of everything from soil microfauna to higher plants.

Under very acidic conditions, there is considerable solubilization of aluminium in the soil, which becomes a component of potential acidity and, consequently, because it is a phytotoxic element, one of the factors responsible for the unfavourable effects of acidity on plants.

In general, the harmful effects of aluminium on plants are mainly reflected in the roots. According to Aguiar (1998), Al^{+3} toxicity reduces cell division, prevents the production of growth regulators in the new parts of the roots, making cell division and the absorption of Ca^{+2} impossible, which only occurs in the apical region of the root. Thus, the roots become slower to elongate, thickening and not branching normally. Adventitious roots proliferate as long as the crown of the plant is alive. In more advanced stages of toxicity, the aerial part is also damaged, and there is a close correlation between the weight of the roots and the aerial part of the plants.

In extreme conditions, in soils with free acids, organic acids or sulphuric acid, in addition to aluminium, iron and manganese are solubilized (pH below 4.5), which are also toxic to plants if absorbed in excessive quantities.

According to Raij (1989), aluminum is the main toxic element associated with soil acidity. Its action is felt in the roots of plants, unlike manganese, another element that can occur in toxic levels in soils, whose effect is more important in the aerial part. According to Primavesi (1990), if exchangeable aluminum does not exceed 40% of the cations in the effective CEC, it is generally not harmful to plants.

In addition to the solubilization of aluminium, iron and manganese, in acidic soils there is a reduction in the mineralization of organic matter, reducing the release of nutrients, especially nitrogen, into the soil solution.

2.3- Soil acidity

2.3.1- Origin of Soil Acidity

Soils can be naturally acidic due to the scarcity of bases in their parent materials, or due to pedogenetic processes that favor the removal of bases from soils. Once formed, over time, soils can become acidic or have their acidity increased due to the extraction of basic cations by crops and, or, the leaching of these cations since acidification consists of the removal of the bases calcium, magnesium, potassium and sodium from soils by percolating water.

In tropical regions with abundant rainfall, the leaching of exchangeable bases from soils is intense, increasing the concentration of hydrogen ions in the medium and thus increasing the acidity of soils.

For soil acidification to occur due to base leaching, co-ions must be present in the soil solution. Co-ions are ions of the same charge as the soil particles that originate from the anions of excess salts in the soil solution. They move through the downward flow of water, binding to the cations present in the soil solution through charge neutralization (Raij, 1991).

The nitrification process consists of the transformation of ammonium ion into nitrate ion, represented by equation 1. Its contribution to soil acidification is due to the release of two hydrogen ions into the soil solution which, because they have a high affinity with the soil, react with it, occupying exchange positions and releasing a corresponding cation.

$$NH_4^+ + 2O_2 \rightarrow NO_3^- + 2H^+ + H_2O \qquad (1)$$

This acid reaction manifests itself both for nitrogen fertilizers added to the soil and for the final mineralization processes of soil organic matter and/or plant residues.

2.3.2- Nature of Soil Acidity

Soil acidity can be classified into exchangeable and non-exchangeable potential acidity and active acidity. Exchangeable potential acidity corresponds to the H + Al adsorbed in the exchange complex, i.e. on the surface of the clays; non-exchangeable potential acidity corresponds to the H^+ that is part of the structure of the clays or the acid groups in the organic matter. It can be determined using different extractant solutions. However, when assessing soil fertility, H+Al is usually extracted with 0.5M calcium acetate solution at pH 7.0 and exchangeable Al is extracted from the soil with 1M potassium chloride solution. The active acidity, which corresponds to the free H^+ ion in the soil solution, is assessed by measuring the soil pH, i.e. determining the H ion activity in a suspension of soil and water.

2.3.3- Soil Acidity Correctives

Products capable of neutralizing soil acidity, known as soil acidity correctors, are those capable of neutralizing

H^+ and Al^{3+} , which is done by the anion OH and, or, HCO_3^- , generated by their basic components, and also have plant nutrients in their composition, mainly calcium and magnesium (Alcarde, 1986). In this way, these products are able to raise the pH of the soil to a satisfactory level in order to promote better crop development; correct calcium and magnesium deficiencies; eliminate exchangeable aluminum toxicity and improve the physical environment for the roots.

According to Brazilian legislation, acidity correctives are classified as: limestone, agricultural quicklime, agricultural hydrated lime (or extinct lime), calcined limestone, basic steel slag and calcium carbonate, with calcium carbonate ($CaCO_3$) being considered the reference standard.

Limestone is obtained by grinding limestone. Its constituents are calcium carbonate ($CaCO_3$) and magnesium carbonate ($MgCO_3$). Depending on their $MgCO_3$ content, limestones can be classified as: calcitic, with a low MgCO3 content (less than 10%); magnesian, with a medium MgCO3 content (10% to 25%); and dolomitic, with a relatively high MgCO3 content (more than 25%). Depending on their geological nature, limestones are also classified as sedimentary and metamorphic.

Agricultural quicklime, agricultural hydrated lime and calcined limestone are products obtained industrially by the calcination or complete burning of limestone, the hydration of quicklime and the partial calcination of limestone, respectively. Basic steel slag is a by-product of the iron and steel industry and calcium carbonate is obtained by grinding marl (terrestrial deposits of calcium carbonate), coral and sambaquis (marine deposits of calcium carbonate).

The constituents of these products are calcium oxide (CaO) and magnesium oxide (MgO), calcium hydroxide ($Ca(OH)_2$) and magnesium hydroxide ($Mg(OH)_2$), and calcium silicate ($CaSiO_3$) and magnesium silicate ($MgSiO_3$). These give rise to the anions CO_3^{2-} , OH^- and SiO_3^{2-} which are responsible for neutralizing the ions H^+ and Al^3 +, i.e. the acidity of the soil. Carbonates and silicates are weak bases and therefore have a weaker neutralizing action, while hydroxide, being a strong base, has a faster and more energetic action in neutralizing acidity. The elements calcium and magnesium are not neutralizers, but plant nutrients.

The index that indicates the potential or theoretical capacity of the soil improver to neutralize soil acidity is the Neutralizing Power (NP). This depends not only on the content of neutralizers in the soil improver, but also on the chemical nature of the neutralizer.

The neutralizing power of an acidity corrector is determined analytically by reacting a sample of the corrector with a known quantity and an excess of relatively dilute (0.5M) hot hydrochloric acid, allowing the corrector to exert its ability to neutralize the acid. According to the principle of chemical equivalence, the amount of acid neutralized is equivalent to the amount of neutralizing constituent contained in the sample. By the same principle, whatever the constituent of the sample, it is considered to be $CaCO_3$, which is taken as the standard. Thus, the amount of CaCO3 that should be in the sample is calculated and the result is expressed as the "equivalent percentage of calcium carbonate" (%$CaCO_3$).

The determination of PN only indicates whether the product is alkaline. It does not make it possible to characterize the chemical nature of the neutralizer, i.e. whether it is carbonate, oxide, hydroxide or silicate, nor whether the product is a soil acidity corrector, i.e. whether the base is associated with calcium or magnesium.

Brazilian legislation requires minimum PN values for correctives: 67%EcaCO3 for limestone, 125%EcaCO3 for virgin agricultural limestone, 94%EcaCO3 for hydrated agricultural limestone, 80%ECaCO3 for calcined agricultural limestone.

2.3.4- Neutralizing Soil Acidity

Due to the problems caused by the acidity of soils, mentioned above, their neutralization is necessary.

Limestone is the product most commonly used to neutralize soil acidity. Calcium carbonate, which is part of its composition, is a salt with very low solubility, but the small amount of dissociation that occurs in the presence of carbon dioxide is enough to trigger a series of reactions that result in the neutralization of soil acidity.

The first reaction is the hydrolysis of calcium carbonate, represented by the following reaction:

$$CaCO_3 + H_2O \longrightarrow Ca^{2+} + HCO_3^- + OH^- \quad (2)$$

This hydrolysis forms OH^-, which neutralizes the H^+ in the soil solution:

$$H^+ + OH^- \longrightarrow H_2O \quad (3)$$

Bicarbonate ions also react with H^+ in the solution:

$$H^+ + HCO_3^- \longrightarrow H_2CO_3 \quad (4)$$

The carbonic acid decomposes:

$$H_2CO_3 \longrightarrow H_2O + CO_2 \quad (5)$$

Therefore, hydroxyl and bicarbonate react with ion H^+ in the soil solution, resulting in water and CO_2, which is released, respectively.

Aluminium, Al^{3+}, reacts with hydroxyl, OH^-, and is neutralized in insoluble form, precipitating as aluminium hydroxide.

The speed of the lime reaction depends on the acidity of the soil, the particle size of the soil improver and the degree of intimacy between the lime and the soil.

6.3.5- Assessing the need for liming

The need for liming is the amount of corrective agent needed to neutralize the acidity of the soil, from an initial condition to a desired condition, depending on the objective to be achieved, the method used and the behaviour of the crops in the face of acidity.

The best method for assessing the need for liming, which is considered the standard, consists of incubating various soil samples with different amounts of calcium carbonate. After the neutralization reaction has been completed, the pH of each sample is measured and a graph is drawn up using the data, relating pH values and doses of calcium carbonate. From this graph it is possible to establish the amount of calcium required to raise the pH of the soil to any desired value. As this method cannot be used routinely because it is time-consuming, the need for liming is determined using faster methods, which in Brazil are basically three.

1. **Exchangeable aluminum method**

It is based on the idea that aluminum is the main unfavorable factor related to acidity in tropical soils. It is suitable for situations where calcium consumption per unit area is low. However, this method is inaccurate because it does not take into account other sources of soil acidity.

According to Bull and Cantarella (1993), experimental results have shown that aluminium content is not, in many soils, the only component of acidity that affects crop development and that higher yields could be obtained with doses of limestone beyond those needed to neutralize this element.

2. **SMP buffer method:**

With this method, the pH determined in the soil suspension with the SMP buffer solution, or pHsMP, makes it possible to directly establish the amounts of limestone to be applied to the soil by knowing a mathematical model, which relates the pHsMP to the amount of corrective agent to be applied to the soil to raise the pH to 6.0 and/or 6.5.

3. **Base saturation method:**

It is based on raising base saturation to pre-established values for different crops. It is necessary to know the potential acidity and sum of bases of the soil in question.

This method has a great advantage over others due to the ease of calculation and the adaptability to different crops, according to their requirements.

According to Bull and Cantarella (1993), using the SMP cap method raises the soil pH to 6.0 for crops such as corn. The liming requirements for this criterion are much higher than for the aluminum method, but the yields obtained are also higher and economically worthwhile. In Sao Paulo, studies showed that the exchangeable aluminum criterion underestimated the need for lime for corn and other crops. Based on this and other data, Quaggio *et al.* (1983) and Raij *et al.* (1993) proposed the adoption of the base saturation (V) criterion for liming, with 60% being the minimum value of soil CEC saturation to be maintained in order to obtain high yields with maize.

Crops' responses to liming vary according to their genetic makeup, which gives them greater or lesser tolerance to acidic conditions, which is why there are big differences between them. In the case of acidity-tolerant plant varieties, the response to liming may not be accentuated, unlike those that are sensitive to soil acidity. This difference in the behaviour of species and varieties in the face of acidity does not allow for easy generalizations and makes it difficult to establish suitable pH ranges for crops (Raij, 1991).

This fact must be taken into account when rationalizing the use of limestone, as the most economical dose of this material basically depends on the magnitude of the crop's response to liming (Chaves, 1997).

6.3.6- Effects of liming

The incorporation of limestone into the soil, an agricultural practice known as liming, should be done as deeply as possible in order to neutralize the acidity of a larger volume of soil, as limestone has low solubility in the short term and its action will manifest itself mainly in the region of contact with the product.

This is especially important in soils with acidity in the subsurface, where aluminum inhibits root growth (Ernani and Barber, 1991; Miyazawa *et al.*, 1992; Rheinheimer *et al.*, 1994), predisposing the plant to greater damage in the event of a summer drought, reducing nutrient absorption and water absorption (Ernani and Barber, 1991; Ernani *et al.*, 1998). Phosphorus is the nutrient most affected because, due to its low mobility in the soil towards the roots, its absorption depends on a large root system. In the studies carried out by Fernandes *et al.* (2000), it was observed that liming reduced or eliminated the effects of aluminum on phosphorus absorption by bean plants.

According to Haynes (1982) cited by ERNANI *et al.* (1996), increasing the pH of the soil can increase, decrease or even not affect the availability of phosphorus. However, other authors, such as Anjos *et al.* (1981) and Ernani *et al.* (1996) state that increasing the pH increases the availability of this element as it decreases its retention in the soil. Fernandes *et al.* (2000) observed that the concentrations of phosphorus fractions in the plant tissues of bean plants increased in the presence of liming, which was attributed to the lower adsorption capacity of phosphorus in the soil and, consequently, greater availability to plants.

Aluminum toxicity also impairs the translocation of phosphorus from the roots to the aerial part due to the formation of insoluble compounds between these two elements in the root system (Rheinheimer *et al.*, 1994; Braccini *et al.*, 1998; McCormick and Borden, 1974, cited by ERNANI *et al.*, 2000).

With liming and when the soil pH reaches values above 5.5, the aluminum precipitates completely and no longer harms the plants (Ernani and Almeida, 1986) and the concentration of phosphorus in the soil solution increases (Ernani *et al.*, 1996). According to Ernani *et al.* (2000), the combination of phosphorus and limestone applied to the soil increased maize yields, which responded better to phosphate fertilization than to liming.

Kabeya *et al.* (1996) observed that liming favored the accumulation of phosphorus in the soil, reducing the critical level of the element in the tissues of forage plants and grasses. There was an increase in the efficiency of use of this element, which could be seen in the increase in dry matter of the grasses.

In addition to neutralizing aluminium and manganese, liming alters the availability of nutrients in the soil for plants, providing calcium and magnesium as nutrients, increasing the availability of molybdenum and decreasing that of other micronutrients, as observed by Barbosa Filho *et al.* (1992) working with rice. These authors found that there was a decrease in the absorption of Cu, Fe, Mn and Zn by plants as a result of liming.

Forestieri and De-Polli (1990) found satisfactory results for the effects of liming on micronutrients, showing that liming increased the plant's assimilation of molybdenum, as well as finding a highly significant response for sulphur in relation to the weight and total N of the aerial part of maize.

According to Raij *et al.* (1994), the availability of potassium in soils decreases as a result of liming; however, Quaggio *et al.* (1982) observed that there was a reduction in losses of the element through leaching as a result of liming. This effect, according to the authors, may be associated with the increase in pH-dependent negative charges caused by liming.

Some authors have observed that crops such as corn and soybeans have shown marked responses to liming, with a significant increase in productivity (Lantmann *et al.*, 1989; Ernani *et al.*, 1998; Raij, *et al.*, 1994; Miranda and Miranda, 2000). In addition to the increase in calcium and magnesium levels in the soil, Miranda and Miranda (2000) found an increase in these elements in the plants' leaf tissues.

Morelli *et al.* (1992), studying changes in the chemical characteristics of a sandy, acidic soil as a result of increasing doses of limestone, observed increases in calcium content and base saturation in the profile, as well as in sugarcane stalk production. Increases in pH, calcium and magnesium content and base saturation as a result of liming were also observed by Caires *et al.* (2000) in a dark red tarsi soil.

According to Azevedo *et al.* (1996) liming in LATOSSOLO BRUNO favored an increase in pH, calcium, magnesium, sodium and cation exchange capacity, and a decrease in organic carbon, exchangeable aluminum, exchangeable potassium and available manganese. The increase in the level of touchable sodium was due to its presence in the limestone, while the decrease in potassium resulted from its displacement as a result of exchange for calcium, leaching and consumption by the plants. The decrease in extractable manganese was due to its solubility decreasing as the pH rose.

The pH has a great influence on the microflora and microfauna present in the soil and their activity. At pH values lower than 5.5, the activity of bacteria and actinomycetes is low, as they thrive best under neutral conditions. Fungi, on the other hand, are generally better adapted and can thrive over a wider pH range. The biological processes in the soil that occur due to the presence of these microorganisms are influenced by pH. Nitrification and nitrogen fixation occur best under neutral conditions, since the participation of bacteria in this process is decisive. The mineralization of organic matter is also influenced by the reaction of the soil. The speed of ammonification and mineralization of sulphates and phosphates is proportional to pH, and these processes occur better under neutral conditions.

According to Caires *et al.* (1998), liming has beneficial effects on the biological process of N fixation associated with greater availability of organic matter and a reduction in soil acidity. Despite this, no significant changes were observed in the N content of soybean leaves with doses of limestone incorporated into the soil.

According to Chaves (1997), the stability of soil aggregates indirectly influences plant development, and in the same way it is influenced by pH, through the ionic coverage of the exchange complex. Normally, clay flocculation occurs and dispersion can occur when limestone is applied, as observed by Albuquerque *et al.* (2000) in an acidic Latosol. Jorge *et al.* (1991) also observed that the application of limestone to dark red latosols reduced the stability of aggregates larger than 2 and 4 mm. With an increase in soil pH and calcium content, there was a reduction in the content of exchangeable aluminum, which is a cation with greater flocculating power than the previous one.

2.4- Industrialization of Granite

Granite is an igneous, intrusive, crystalline rock with a granular texture, made up essentially of quartz and feldspar, with small amounts of other materials. Granite is basically made up of silica (around 75%) and is generally considered to be a silicate rock.

From a commercial point of view, granite is any non-calcareous rock that can be polished and used as a cladding material. This definition includes rocks of different compositions, even without affinities with granite, but which meet the necessary conditions for use in civil construction, such as granite itself (NITES, 1992, quoted by DA SILVA, 1998).

Industrialization is made up of several phases, which can be summarized as follows: research and extraction studies are carried out to obtain the blocks; these blocks are transformed into slabs in sawmills, constituting the splitting or sawing process, and are subjected to finishing processes to obtain the final product (Silva, 1987).

Exploration work in quarries is very simple and low-cost, limited to taking surface samples, without the need to drill boreholes, wells or galleries for analysis. In the massifs, the blocks are obtained through the use of explosives and are squared off by the continuous drilling of pneumatic jackhammers, which are blowtorches that, using fuel and compressed air, generate high-impact flames for cutting granite. In the matacôes, also through hammer holes, the blocks are outlined and obtained by abrasives, but the shape of the block is irregular and its size is subject to the size of the matacão (Romagnoli, 1985).

The exploitation of ornamental rocks is made possible by the large quantities of granite that can be mined in the form of "matacôes", which are large masses of granite detached from the primary rock and left loose on the surface, or in the form of "massifs", which differ from the former both because they are or are not at surface level and because they are medium to large in size; in addition, labor is cheap.

After being extracted, the blocks are transported to sawmills for splitting or sawing, the process of transforming the blocks into semi-finished slabs or plates with thicknesses ranging from 1 to 3 cm, using machines called looms, the most common of which use abrasive mixtures.

Looms are sturdy pieces of equipment made up of four columns, which support a frame that makes a pendulum movement. On this frame, steel blades are arranged in the longitudinal direction of the equipment, parallel to each other. These blades are made of carbon steel with high hardness and good mechanical resistance, to better withstand the forces of traction and abrasion (SENAI, 1992, quoted by DA SILVA, 1998).

Nowadays, there is a more widespread use of multiple blade looms (up to 100 or more blades per loom) using an abrasive mixture. The products obtained from this stage of industrialization are semi-finished sawn slabs, using the following main inputs: abrasive component (steel or cast iron grit for sawing the blocks), water, lime (limestone, calcium carbide or other substitute) and electricity.

Steel or cast iron shot are spherical or angular abrasive particles produced from the materials mentioned above. They have a defined grain size and may or may not be heat-treated. This abrasive is the main friction element for cutting granite, being dispersed in the abrasive blade, receiving the impact produced by it and causing the granite particles to break up, thus allowing the block to be split into semi-finished slabs. As they wear down during the process, those with the smallest diameter are often removed and new grit added.

Due to the variety of material being sawn by companies, each one uses a different composition of different grit sizes, or uses spherical and angular grit simultaneously, as well as the option of using cast iron grit. The angular grit has greater cutting power and the spherical grit has the function of reducing imperfections in the shape of the sheet at the time of cutting, providing a better finish.

The abrasive mixture is controlled according to analyses that measure its specific weight, its viscosity and the percentage of active grit, inactive grit and grit dust. Active grit has a higher cutting power than inactive grit, with the latter having minimal cutting power, and its main function is to control the viscosity of the mixture. The only function of the grit powder is to control the viscosity, which is done every two hours.

The abrasive slurry is pumped from a pit and transported through a rubber tube reinforced with steel springs, which prevent the shear that the grit can cause during the flow. Once pumped, it is directed to a central shower and distributed to side showers, being thrown on top of the granite blocks, starting the sawing process. Once this process is complete, the pulp returns to the pit, from where it is pumped out again, closing the cycle. Part of it becomes fine due to the decreasing size of the grit used in this process, and needs to be discarded so as not to compromise the efficiency of the granite sawing.

It is during this phase of stone processing that the waste from sawing granite blocks is generated.

There are also diamond blade looms and those that use diamond discs (block hoists). Diamond blade looms use a water shower to clean, lubricate and cool the blades, which differ from high-hardness carbon steel blades in that they have industrial diamonds attached to the underside. This process is expensive due to the need to replace and maintain the blades, as well as the high cost of purchasing the loom. This process has higher unit processing costs than abrasive blade cutting, and is commonly used for small blocks (no more than 1m high) and the production of small, standardized pieces (FIEC, 1992, quoted by DA SILVA, 1998).

Because no abrasive mixture is used in this loom, the amount of waste generated is smaller and is therefore considered to have less impact on the environment.

2.5- Breakdown of Granite Blocks at Indùstria de Beneficiamento de Granitos do Brasil S/A - POLIGRAN

The state of Paraiba has 144,874 m^3 of granite reserves (Anuârio Mineral Brasileiro, 1998), making it possible to exploit blocks for industrial processing. In industry, each cubic meter of granite block produces 34m^2 of polished slabs.

POLIGRAN uses granite from its own quarries located in municipalities in the state of Paraíba.

To start cutting the blocks, we usually start with a mixture from a previous sawing. When this is not possible, a mixture is made with the following composition: 50 to 100 kg of lime or 30 to 50 kg of calcium carbide sludge (CaC_2), with the grit being added continuously in small quantities dosed manually or by an automation mechanism (SENAI, 1995, cited by DA SILVA, 1998).

The abrasive mixture is pumped over the block(s) evenly and continuously, on all blades. In general, these systems are made up of multi-arm distributors, each with nozzle points for spraying the abrasive fluid.

During the sawing of granite blocks, the composition of the abrasive slurry on the looms varies greatly over a short period of time. This is due to the addition of rock fragments (dust), metallic residues from the grit and the blades to the abrasive mixture which, having had its initial basic composition altered, gradually loses efficiency.

If these fragments are not removed periodically, the mixture will gather more and more solids, considerably increasing its viscosity, thus reducing its cutting power, to the point that, as the percentage of these solid particles increases, much of the energy applied between the blade and the rock will be wasted by crushing fragments of already cut rock, as well as increasing the mechanical wear of the blades and the loom, which can result in defects in the sawn slabs.

It is necessary to control the viscosity of the abrasive mixture by periodically adding water and discarding the portion of the mixture with the smallest granulometry, which has physical characteristics that could hinder the development of the abrasion required for the block cutting process. The discarded waste consists of metal fragments from the wear of the blades and the grit, as well as lime or lime substitutes (limestone or blast furnace slag).

In the water treatment unit (UTA), approximately 98% of the water used to split the rocks is reused. It is pumped to a storage tank where it is treated with aluminum sulphate, which flocculates the suspended particles and makes them decant. After decanting, the water is taken to a tank to be used again to break up the granite blocks.

The amount of waste (sawdust sludge) generated by this activity in the industry is 792 m^3 per year, which corresponds to the production of two multi-blade looms. With the use of a filter press (to be installed), the water will be fully utilized and the waste generated will be a fine powder.

3-MATERIAL AND METHODS

This work, carried out in a greenhouse and in the Irrigation and Salinity Laboratory (LIS) of the Agricultural Engineering Department of the Science and Technology Center of the Federal University of Campina Grande, PB, was conducted in two separate stages. In the first stage, a preliminary study was carried out on the effects of incorporating granite sawdust sludge into the soil. In a second stage, a study similar to the previous one was carried out, plus an evaluation of crop development as a function of granite sawdust sludge.

3.1. Soil

The soils used in this work came from the Agreste da Borborema region in the state of Paraiba. According to EMBRAPA (1999), they were classified as ARGISSOLO VERMELHO-AMARELO Distrófico.

The soil samples used were collected at a depth of 0 - 20 cm, air-dried, crushed and passed through a sieve with a 2 mm mesh opening. Their chemical characteristics were determined according to the EMBRAPA methodology (1997).

3.2. Granite sawdust sludge

The granite sawdust used in the study came from the Polimento de Granitos do Brasil S/A - (POLIGRAN) processing industry, located in the municipality of Campina Grande, Paraiba State.

The granite sawdust sludge samples used in the two stages of the work were collected at different points in the industry's final disposal tank, so that they represented all the material deposited on site, and then mixed to obtain composite samples. After being air-dried and passed through an ABNT 10 sieve, these samples were chemically analyzed and the relative total neutralizing power (PRNT) of each sample was determined.

3.3. Treatments

3.3.1 First stage

The treatments used in the first stage of the work consisted of increasing doses of granite sawdust sludge, 0; 0.4; 0.8; 1.2; 1.6; 2.0; 2.4; 2.8; 3.2; 3.6 and 4.0 gkg-1 of soil, determined on the basis of the amount of calcium carbonate needed to raise the soil's base saturation to 60%, i.e. 0.8 gkg^{-1} of soil. Considering the PRNT of the granite sawdust sludge sample, these doses should raise the soil's base saturation to 30.8; 32.4; 40.7; 44.5; 48.3; 54.1; 55.9; 59.7; 63.5 and 67.3%, respectively.

3.3.2 . Second Stage

The treatments used in the second stage of the work consisted of increasing doses of granite sawdust sludge, i.e. 0; 0.56; 1.2; 1.84; 2.48; 3.12; 3.76; 4.4; 5.04; 5.68 and 6.32 gkg^{-1} , corresponding, respectively, to the amount needed to raise the soil's base saturation to 36.7; 42.5; 48.2; 54.0; 59.8; 65.6; 71.4; 77.2; 83.0 and 88.8%.

3.4. Setting up and conducting experiments

3.4.1 - First Stage (Preliminary Experiment)

Each experimental unit consisted of a 2.0 kilogram sample of soil, placed in a plastic pot with a drain at the bottom, which was mixed and homogenized with the doses of mud determined for each treatment. In order to give the slurry a chance to interact with the soil and thus make better use of its beneficial characteristics, the experimental units were incubated for 70 days, with soil humidity maintained at close to 80% of field capacity. The distilled water lost during this period was replaced by weighing the pots daily.

In order to assess the variation in soil pH during the incubation period, every 10 days from the start of the experiment, 0.05 kg of soil was removed from each experimental unit, dried in the air, passed through a 2 mm sieve (ABNT 10) and submitted for pH determination in water (1:2.5).

At the end of the incubation period, the soil samples were chemically characterized according to EMBRAPA (1997).

3.4.2 Second stage

The experiment in the second stage of the work was set up in the same way as the previous one, with each experimental unit consisting of a 2.5 kilogram sample of soil, which was incubated for 180 days.

After the incubation period, soil samples of approximately 0.05 kilograms were taken from the experimental units, dried in the air, passed through a 2.0 mm sieve (ABNT 10) and subjected to chemical analysis according to the methods described by EMBRAPA (1997).

At the end of the experiment, the soils from the experimental units were subjected to chemical analysis.

3.4.2.1 Installation and Conduct of the Test Crop

Taking into account the results of the chemical analysis of the soil samples as well as the nutritional needs of the maize (*Zea mays* L) used as a test plant, after the incubation period, all the experimental units received phosphate fertilizer. The fertilizer used was simple superphosphate (18% P_2O_5, 20.6% Ca and 12% S), previously macerated in a mortar and passed through a 0.50 mm sieve (ABNT 35) to allow for better homogenization with the soil.

In order to guarantee the amount of available phosphorus needed for the plants to grow, the amount of fertilizer each experimental unit received corresponded to four times the calculated dose. After fertilization, five maize seeds were sown in each experimental unit and, fifteen days after sowing, two plants were left per unit.

Soil moisture was maintained at close to 80% of field capacity by daily weighing of the pots (Bataglia 1990).

The experiment was completed 60 days after sowing. The plants were cut close to the soil surface and the roots were washed to separate them from the soil.

3.4.2.2 Plant Characteristics Evaluated

After cutting, the plants were immediately weighed to obtain their fresh biomass. The length of the leaves was determined and the leaf area was determined using a LI-COR electronic meter, model LI- 3000. The aerial and root biomass was then dried in a hot air forced circulation oven at 60° C until constant weight, obtaining the dry biomass of the aerial part and the dry biomass of the roots. The total dry biomass of each plant was then obtained. The material was weighed on a 0.0001g precision scale. The aerial part/root ratio, which allows us to say which of the two parts of the plant was most affected by the soil conditions, was obtained from the quotient between the weights of the dry biomass of the respective parts. The leaf area ratio, a measure of the relative size of the assimilation apparatus which serves as an appropriate parameter for evaluating the effects of genotypes, climate and the management of plant communities (Ferri, 1985), was obtained by dividing the leaf area by the total dry biomass.

All the dry biomass was passed through a mill for subsequent analysis of the macronutrients nitrogen (N), phosphorus (P) and potassium (K) (Tedesco, 1995). Each repetition was represented by the average of two plants.

3.5. Experimental Design

The experiments were set up in a completely randomized design, consisting of 11 treatments with 3 replications.

Multiple regression analysis was carried out using the Stepwise method for the results relating to the yield of fresh and dry biomass of the aerial part and the dry biomass of the roots, as well as regression analysis of the analyzed soil parameters according to the methodology adopted by John *et al.* (1994). The F-test was used to determine the level of significance of the variables studied. When the regression deviation is statistically significant, we conclude that there is at least one polynomial component above the third degree that is statistically significant. In this study, we only considered models that were significant up to the third degree since, from a practical point of view, models with a higher degree do not serve to explain the behavior of the variables studied.

3.6. Description of the Methodologies Adopted

3.6.1 Soil

The exchangeable cations calcium, magnesium and aluminum were extracted with 1M KCl. Sodium and potassium were extracted using a solution of 0.05M HCl and 0.0125M H2SO4 and potential acidity (H+Al) using 0.5M calcium acetate pH 7.0 (EMBRAPA, 1997). The cation exchange capacity was estimated from these results. The pH was determined using the 1:2.5 ratio of soil to water. A 0.2M potassium bichromate solution was used to determine the organic matter content. Electrical conductivity was analyzed according to the methodology recommended by Richards (1954), cited by EMBRAPA (1997).

3.6.2 Granite Sawdust Sludge

The granite sawdust slurry was chemically characterized using the methods described above for soils. It was also analyzed using the methodology adopted by EMBRAPA (1979) for the analysis of limestone, where concentrated hydrochloric acid was used to react with the limestone in the material.

The mineralogical characterization of the granite sawdust sludge used in the second stage of the work was carried out using fluoridated acid leaching.

3.6.3 Dry biomass

The macronutrients present in the plant tissues were analyzed according to the methodology proposed by Tedesco (1995).

Phosphorus levels were determined by spectrophotometry on an aliquot of the extract after adding ammonium molybdate and aminonaphtholsulfonic acid. Potassium levels were determined by flame photometry after diluting the extract, adjusting the sensitivity of the apparatus with the appropriate standards. Total nitrogen was analyzed using the Kjeldahl method. In this method, the nitrogen in the sample is converted to the ammoniacal form (NH_4^+ - N) by digestion with concentrated sulphuric acid, under heat, with the addition of oxygenated water.

4-Results and discussion

Tables 1 and 2 show the results obtained from the chemical analysis of the soil samples and the granite sawdust sludge used in the first and second stages, respectively. Table 3 shows the results of the mineralogical composition of the mud used in the second stage of the work.

Table 1. Chemical characteristics of the soil samples, 1 and 2, used in the first and second stages of the work, respectively.

Features	Soil sample 1	Soil sample 2
pH H_2O (1:2.5)	4,75	4,37
EC $(dsm)^{-1}$	0,09	0,20
Organic matter $(gkg)^{-1}$	7,4	30,3
Phosphorus $(mgdm)^{-3}$	3,1	1,4
sodium $(cmolc.kg)^{-1}$	0,04	0,16
Potassium $(cmolc.kg)^{-1}$	0,33	0,18
Magnesium $(cmolc.kg)^{-1}$	0,99	1,41
Calcium $(cmolc.kg)^{-1}$	0,55	1,22
Hydrogen $(cmolc.kg)^{-1}$	4,09	5,87
Aluminum $(cmolc.kg)^{-1}$	0,50	0,52
V (%)	29,38	31,73

Table 2. Chemical characteristics of the mud samples, 1 and 2, used in the first and second stages of the work, respectively.

Features	Mud sample 1	Mud sample 2
pH H_2O (1:2.5)	12,68	9,68
EC $(dS\ m)^{-1}$	5,36	2,26
Organic matter (%)	-	-
Phosphorus $(mg.dm)^{-3}$	-	-
Sodium $(cmolc.kg)^{-1}$	3,24	5,86
Potassium $(cmolc.kg)^{-1}$	17,9	2,15
Magnesium $(cmolc.kg)^{-1}$	1,84	3,63
Calcium $(cmolc.kg)^{-1}$	39,44	3,49
Hydrogen $(cmolc.kg)^{-1}$	-	-
Aluminum $(cmolc.kg)^{-1}$	-	-
$CaCO_3$ (%)	12,5	15,00
$MgCO_3$ (%)	10,5	21,84
CaO (%)	7,0	8,40
MgO (%)	5,0	10,53
PRNT (%)	24,48	33,47

Table 3. Mineralogical characteristics of the granite sawdust sludge sample used in the second stage of the work.

Features	Value (%)
Loss on redemption (Pr)	4,80
Silicon oxide (SiO_2)	57,30
Insoluble residue (IR)	2,27
Iron oxide (Fe_2O_3)	8,00
Aluminum oxide (Al_2O_3)	15,00
Calcium oxide (CaO)	5,00
Magnesium oxide (MgO)	Traits
Sodium oxide (Na_2O)	3,38
Potassium oxide (K_2O)	3,63

3.7. First Stage (Preliminary Experiment)

The variations in pH $_{H2O}$ (1:2.5) of the experimental units as a function of the treatments in the seven sampling periods are shown in Figure 1. It can be seen from the trend of the curves that the variations in all the samples had an increasing behavior, however, in each experimental unit, the pH decreased over time, in agreement with what was observed by Anjos *et al.* (1981). According to Quaggio (1986), this decrease in pH values over the incubation period may be due to the accumulation of salts from the mineralization of organic matter.

Figure 2 shows the variation in the base saturation percentage of the experimental units, in the last sample, as a function of the treatments. Considering the high pH value of the granite sawdust sludge and the base saturation percentages predicted depending on the treatments, the results were lower than expected. There was little variation in the pH values of the experimental units and the application of the highest dose of granite sawdust sludge caused the soil's base saturation to be around 49%. This can probably be explained by the low levels of neutralizing materials found in the granite sawdust sludge and, consequently, its low PRNT value (Table 2). Results similar to these were obtained by Costa *et al.* (1995) working with waste from the marble and granite processing industry in the state of Espirito Santo.

The increasing behavior of the pH values and the percentage of base saturation as a function of the treatments, according to the trend of the curves, and the need to learn more about this behavior, were the reasons for setting up a second experiment with higher doses of granite sawdust sludge.

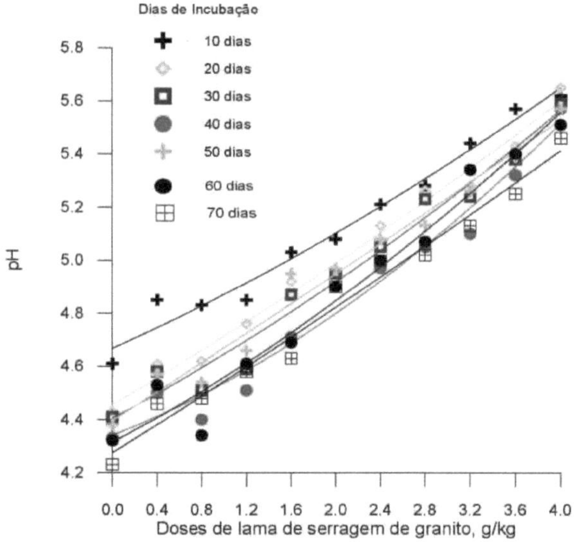

Figura 1. Variation in pH as a function of doses of granite sawdust sludge at different sampling times.

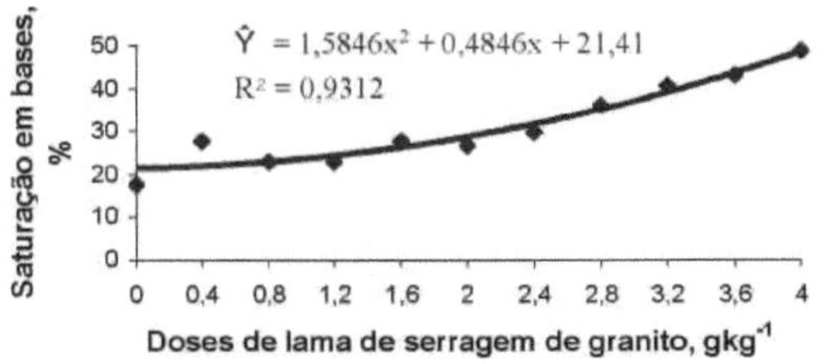

Figura 2. Variation in the percentage of base saturation of the experimental units as a function of the doses of granite sawdust.

Table 4 shows the values of the chemical characteristics of the experimental units as a function of the increasing doses of granite sawdust sludge (treatments) after the incubation period. Due to the presence of calcium in the granite sawdust sludge and its ability to neutralize acidity, the calcium content increased and the potential acidity values (H+Al) decreased. The contents of the other elements and the organic matter values hardly varied when compared to the control.

Table 4. Chemical characteristics of the experimental units as a function of the treatments, after the incubation period (70 days).

Treatments	MO	Ca	Mg	In	K	H+Al	P
gkg 1-	gkg 1-	cmolckg -1----------------					mgdm-3
0	17,2	0,43	0,75	0,075	0,07	6,23	0,30
0,4	16,6	0,65	0,79	0,065	0,07	4,12	0,30
0,8	20,2	0,63	0,76	0,070	0,07	5,15	0,30
1,2	21,0	0,71	0,80	0,085	0,08	5,67	0,25
1,6	15,4	0,89	0,83	0,060	0,08	4,86	0,35
2,0	15,9	0,82	0,69	0,075	0,07	4,54	0,30
2,4	13,1	0,93	0,65	0,075	0,07	4,06	0,25
2,8	14,2	1,13	0,75	0,090	0,08	3,62	0,40
3,2	13,0	1,14	0,77	0,050	0,08	2,96	0,30
3,6	13,6	1,22	0,62	0,075	0,08	2,64	0,40
4,0	15,8	1,34	1,12	0,095	0,09	2,78	0,40

3.8. Second stage

Table 5 shows the values of the chemical characteristics of the experimental units as a function of the increasing doses of granite sawdust sludge (treatments) after the incubation period.

Table 5. Chemical characteristics of experimental units as a function of treatment, after the incubation period (180 days).

Treatment	pH (H_2O)	C.E.	M.O.	In	K	Ca	Mg	Al	H	H+Al	P
gkg-1		dSm-1	g.kg-1	cmolc.kg-1							mgdm-3
0	4,50	0,15	29,3	0,07	0,19	1,09	1,34	0,60	6,71	7,31	1,7
0,56	5,03	0,13	36,2	0,08	0,14	1,79	2,38	0,05	4,95	5,00	1,8
1,2	5,07	0,15	32,1	0,08	0,15	1,58	2,21	0,12	5,44	5,56	1,6
1,84	5,08	0,14	32,2	0,09	0,18	1,90	2,01	0,08	5,24	5,32	2,2
2,48	5,10	0,15	31,1	0,08	0,15	1,76	1,91	0,10	5,26	5,36	1,1

3,12	5,06	0,14	29,4	0,08	0,17	1,67	1,92	0,10	5,42	5,52	1,7
3,76	5,14	0,18	38,2	0,10	0,22	1,90	2,21	0,07	5,12	5,19	2,6
4,4	4,94	0,26	38,7	0,15	0,29	1,98	2,41	0,10	6,04	6,14	1,6
5,04	4,90	0,35	43,4	0,18	0,35	2,09	2,71	0,09	6,42	6,51	2,6
5,68	4,73	0,32	39,2	0,18	0,39	1,96	2,50	0,14	7,03	7,17	2,9
6,32	4,71	0,34	41,4	0,19	0,41	1,93	2,37	0,17	7,44	7,61	3,1

3.8.1 Changes in Soil Acidity Caused by Granite Sawdust Sludge After Incubation Period

The variation in pH values in water (pH $_{H2O}$) as a function of the treatments after the incubation period and the regression equation fitted to the curve with its coefficient of determination can be seen in Figure 3.

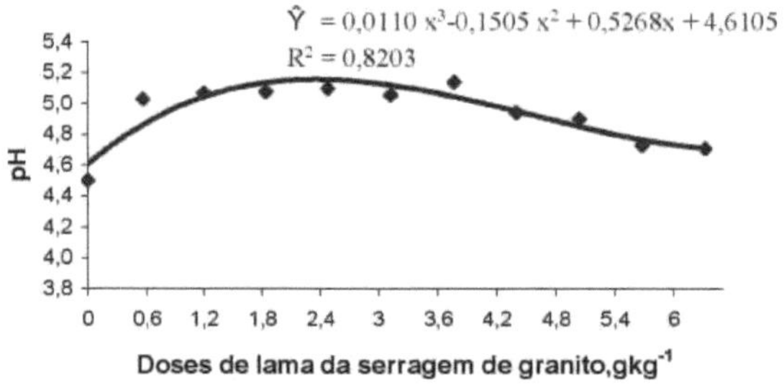

Figura 3. Variation in pH $_{H2O}$ as a function of doses of granite sawdust slurry.

The behavior of the curve shows that, in general, the values increased in relation to the initial pH, in the same way as in the preliminary experiment (Figure 1), decreasing at the higher doses. This decrease may be a consequence of the increase in the saline concentration of the solution due to the increase in sludge doses, which did indeed occur, as shown by the EC results (Table 5). This pH behavior as a function of EC was also observed by Franchini *et al.* (1999) when evaluating chemical changes in acidic soils after the application of plant residues. The variation in EC values according to the treatments and the regression equation to which they fitted significantly at 5% by the F test, as shown by the value of the explanation coefficient, can be seen in Figure 4.

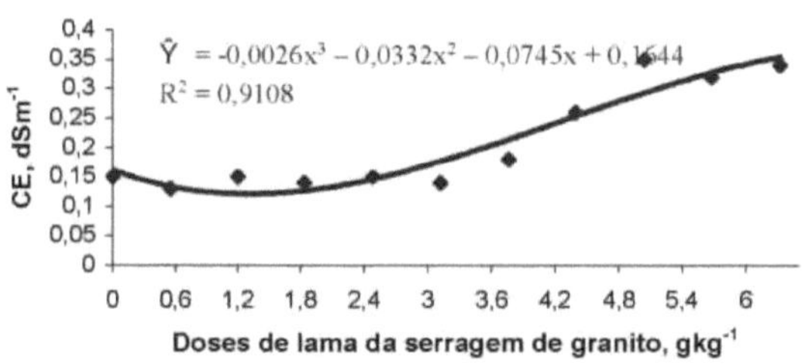

Figura 4. Variation in electrical conductivity (EC) values as a function of doses of granite sawdust.

Although the increase in pH values was smaller and smaller, the application of granite sawdust sludge to the soil caused the pH values of all the experimental units to rise above those of the control. According to the estimated model, the highest pH value obtained was for the 2.36 gkg^{-1} treatment, where the value was 5.16.

The pH values remained in the range that characterizes the soil as strongly acidic, according to EMATER (1979). This is possibly due to the low carbonate content in the granite sawdust sludge and/or the high organic matter content (Table 5) in the soil, which possibly gives it a high buffering power. According to Faquin *et al.* (1998), in soils with high buffering power, even high doses of correctives cause small increases in pH.

Aluminum (Al) content and potential acidity (H+Al) values generally decreased as a result of the treatments (Table 5); the decrease was around 85% and 27%, respectively, considering the first six doses of granite sawdust sludge. The fact that the exchangeable aluminium in the experimental units that received the treatments was not completely neutralized agrees with the observations of Cunha *et al.* (1996). More important than the absolute values are the percentages of aluminum saturation in the experimental units, which also decreased as a result of the treatments, by an average of 88% compared to the control (Table 6).

Soil acidity is neutralized by applying lime to the soil, the recommended doses of which can be calculated using methods based on the exchangeable aluminium content, the exchangeable calcium + magnesium content and the increase in base saturation. Based on these three methods, the quantities of calcium carbonate needed to neutralize the acidity of the experimental units were calculated, even though they had received the treatments (Table 6). When calculating base saturation, a value of 60% was considered ideal.

Although low, the slurry from granite sawing will always have a neutralizing effect on the acidity of the soil, since calcium oxide is usually mixed with the water used to saw the granite blocks.

Table 6. Cation exchange capacity (CEC), sum of bases (S), percentage of base saturation (V), percentage of aluminum saturation (Al) of the experimental units and amount of calcium carbonate recommended by three different methods depending on the treatments, after the incubation period.

Treatments	CTC pH7[1]	Effective CTC[2]	S[3]	V[4]	Al[5]	Calcium carbonate recommendation[6] Method of Al[a]	Ca + Mg method[b]	Method of Saturation bases[c]
$g kg^{-1}$		-cmolckg $^{-1}$--		--%-	-----	-	$t ha^{-1}$ -.	
0	10,00	3,29	2,69	29,6	18,23	0,90	0,57	3,04
0,56	9,39	4,44	4,39	46,75	1,13	0,07	-	1,24
1,2	9,58	4,14	4,02	41,96	2,89	0,18	-	1,73
1,84	9,50	4,26	4,18	44,00	1,88	0,12	-	1,52
2,48	9,26	4,00	3,90	42,12	2,50	0,15	-	1,65
3,12	9,36	3,94	3,84	41,02	2,54	0,15	-	1,78
3,76	9,62	4,50	4,43	46,05	1,55	0,10	-	1,34
4,4	10,97	4,93	4,83	44,03	2,03	0,15	-	1,75
5,04	11,84	5,42	5,33	45,02	1,66	0,13	-	1,77
5,68	12,20	5,17	5,03	41,23	2,70	0,21	-	2,29
6,32	12,51	5,07	4,90	39,17	3,35	0,25	-	2,60

1. CTC= Ca+Mg+Na+K+H+Al; 2. Effective CTC= Al+Ca+Mg+Na+K; 3. S= Ca+Mg+Na+K; 4.V= (S/CTC)x100; 5.Al= (Al/S+Al)x100; 6a. (t/ha = Al x 1.5); 6b. (t/ha= [3- (Ca + Mg)]); 6c (t/ha= [(V2-V1)xCTC] /100)

The results show that, using the method based on the calcium plus magnesium content, the experimental units that received the granite sawdust sludge did not need to receive calcium carbonate; using the method based on the exchangeable aluminum content, these experimental units would have to receive it, but in smaller quantities than that required for the control; this reduction in the quantities of calcium carbonate would vary from 72 to 92%. In relation to the percentage of base saturation method, it can be seen that the recommended quantities of calcium carbonate would also be lower than that corresponding to the control; on average, there would be a reduction of 49%, considering the first six doses of sludge. Although this reduction is apparently significant, it should be borne in mind that the experimental units had already received the treatments. Cunha *et al.* (1996), using waste from the marble and granite industry to correct soil acidity, found similar results when using waste from the granite industry, where it did not completely neutralize the soil's aluminium content.

As can be seen in Table 6, the V% values of the experimental units increased according to the treatments; however, due to the inefficiency of the sludge, only the first treatment achieved the expected V% value of 36.7%.

Considering that granite sawdust sludge has a low acid neutralizing effect, and that its use would help to reduce the environmental impact caused by its disposal and accumulation, it might be advisable to enrich it with limestone. Before doing so, however, studies should be carried out into the ideal quantities of mud and limestone to be mixed and the economic viability of using such a mixture.

3.8.2 Changes in Calcium, Magnesium, Potassium, Sodium and Phosphorus in the Soil Caused by Granite Sawdust Sludge After the Incubation Period

From the adjusted regression equations, curves were obtained (Figure 5) which show that there was an increase in the levels of magnesium, calcium and potassium as a function of the increasing doses of granite sawdust sludge. However, the doses did not provide significant increases in the levels of calcium and magnesium, but were only enough to raise the initial levels (control) of these elements, which were close to the lower limit of the range of values classified as medium (2.1 - 5.0 $cmolckg^{-1}$), according to EMATER (1979), to levels close to the upper limit of this same range, as can be seen in Table 5. In other words, the values continued to be classified as medium.

Although the granite sawdust sludge has potassium in its composition, the first five experimental units that received the treatments did not show increasing potassium levels, probably due to the increase in pH, which according to Raij (1991), can decrease the availability of the element. In the last experimental units, there was an increase in potassium content due to the contribution of the treatments and/or the decrease in pH, but only the last two levels, i.e. those corresponding to the treatments 5.68 and 6.32 gkg^{-1} (Table 5), could be classified as high levels (EMATER, 1979).

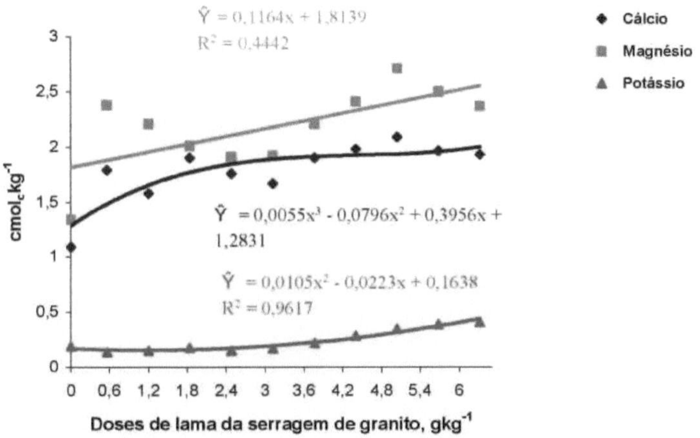

Figura 5. Variation in calcium, magnesium and potassium levels as a function of doses of granite sawdust.

The behavior of the adjusted curve, according to the regression equation presented in Figure 6, shows how the sodium content of the experimental units varied as a function of the increasing doses of granite sawdust sludge, significantly at 1% probability, according to the F test. This variation, i.e. the increase in sodium levels, as also observed by Azevedo *et al.* (1996), was possibly due to the presence of this element in the mud. Even so, the levels observed are considered very low, allowing the experimental units to be classified as normal in relation to the classification of soils in terms of the percentage of exchangeable sodium (PST) (EMBRAPA, 1998).

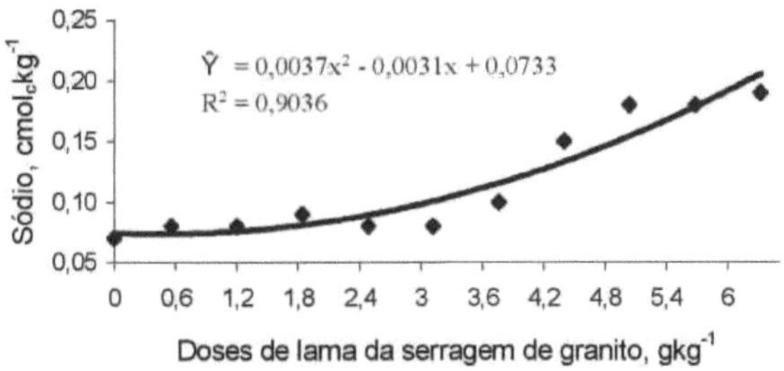

Figura 6. Variation in sodium content as a function of doses of granite sawdust sludge.

The increases in the calcium, magnesium, potassium and sodium levels meant that the sum of the bases (S) increased, as can be seen in Table 6. As a result, the effective CEC of the experimental units also increased, even considering the decrease in aluminum levels. With regard to the CEC at pH 7.0, it can be seen that it decreased in the first six experimental units that received the treatments and then increased, mainly due to the increase in the levels of exchangeable cations and hydrogen. Similarly, the percentage of base saturation (V%) of the experimental units varied according to the treatments. This variation meant that all the experimental units had V% values above that of the control. Despite the increases observed in the variables S, effective CEC, CEC at pH 7.0 and V% in general, they were not enough to change the interpretation of the values previously existing in the soil, which according to Mello *et al.* (1983), were maintained as average values.

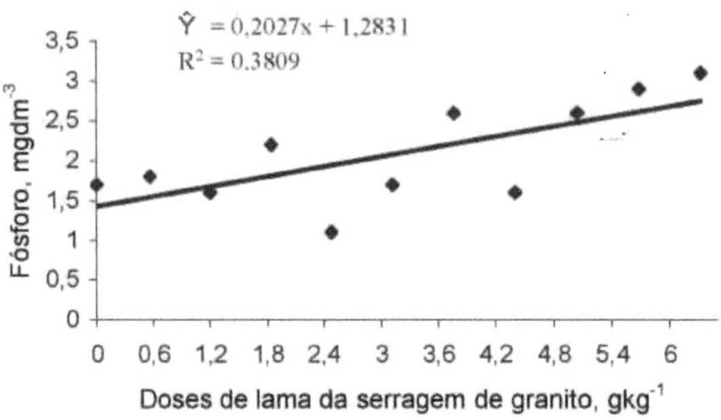

Figura 7. Variation in phosphorus content as a function of doses of granite sawdust.

As the granite sawdust sludge does not have phosphorus in its chemical composition, it can be assumed that the increases observed in the levels of this element as a function of the treatments (Figure 7) were due to the variation in pH in the experimental units, according to Anjos *et al.* (1981) and Ernani *et al.* (1996). Despite these increases, the phosphorus content of all these units continued to be classified as low (EMATER, 1979).

The variation observed in the organic matter content as a function of the treatments may be related to the presence of some organic material in the granite sawdust sludge, as can be seen from the loss on ignition data (Table 3), since this parameter normally indicates the presence of organic substances in the material analyzed.

7.2.3. Changes in Soil Chemical Characteristics After Planting

Table 7 shows the values of the chemical characteristics of the experimental units as a function of the treatments, after the plants had been cut (end of the experiment).

Table 7. Averages of the analytical results of the final soil as a function of the doses of granite sawdust sludge applied to the soil.

Treatments	**pH (H_2O)**	**C.E.**	**M.O.**	**In**	**K**	**Ca**	**Mg**	**Al**	**H**	**H+Al**	**CTC**	**P**
		dSm 1-	**gkg 1-**	**cmolckg^{-1}**								**mgdm^{-3}**
0	4,29	0,22	27,2	0,06	0,04	1,37	1,55	0,65	5,68	6,33	9,35	24,3
0,56	5,01	0,19	33,2	0,08	0,04	2,24	1,78	0,09	4,15	4,24	8,38	16,4
1,2	4,90	0,19	30,8	0,09	0,04	2,08	1,62	0,15	4,58	4,73	8,56	19,4
1,84	4,87	0,22	28,7	0,10	0,04	1,82	1,91	0,13	4,33	4,46	8,33	19,4
2,48	4,89	0,21	25,6	0,09	0,04	1,94	2,15	0,12	4,56	4,68	8,90	17,5
3,12	4,78	0,20	23,6	0,08	0,04	2,03	1,92	0,13	4,77	4,90	8,97	24,9
3,76	4,99	0,19	26,4	0,10	0,05	2,20	2,42	0,08	4,32	4,40	9,17	18,1
4,4	4,83	0,22	30,8	0,10	0,06	2,32	2,32	0,17	4,40	4,57	9,37	18,5
5,04	4,87	0,22	34,0	0,12	0,07	2,32	2,45	0,19	5,42	5,61	10,57	21,7
5,68	4,69	0,21	38,4	0,12	0,07	2,25	2,28	0,28	6,22	6,50	11,22	20,6
6,32	4,60	0,22	38,6	0,11	0,06	2,29	2,20	0,27	6,00	6,27	10,93	25,0

Analyzing these values in relation to those presented in Table 5, i.e. the characteristics of the experimental units before planting, it can be seen that there were some variations, which, in general, will be commented on below.

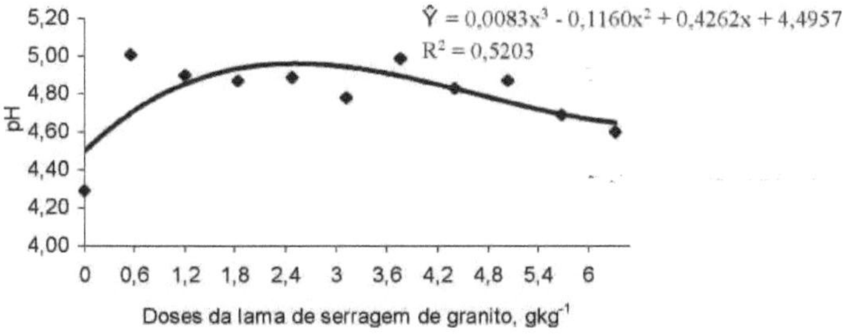

Figura 8. Variation in the pH $_{H2O}$ of the soil at the end of the experiment as a function of the doses of granite sawdust slurry.

The pH values of the experimental units at the end of the experiment were lower than those at the beginning (Table 5), although the behavior of the adjusted curves of the regression equations, shown in Figures 3 and 8, was similar.

The decrease in pH values over time is in line with what was observed in the first stage of this work and also by Escosteguy and Klamt (1998) when incubating soils with ground basalt. Possibly part of this effect can be explained by the reduction in pH also observed in the control, which indicates a process of acidification caused by factor(s) other than the addition of granite sawdust sludge. According to the aforementioned authors, it is possible that the acidification in the control soil was caused by hydrolysis, resulting from the addition of water to the soil, and by the decomposition of organic matter. In the case of the experimental units that received the treatments, during the sludge solubilization process, acidifying compounds may have been formed, or the neutralizing power of the sludge may have been reduced. Similarly, the consumption of bases by the plants and/or the phosphate fertilization of the experimental units may have contributed, in some cases, to the acidification of the soil. According to Malavolta (1985), simple superphosphate fertilizer can slightly lower the soil's pH.

Even though there was a decrease in the pH values, they were still above the initial pH value of the soil, showing that the granite sawdust slurry neutralized part of the soil's acidity (Table 7).

With the exception of the two experimental units that received the highest doses of sludge, in the others there was a decrease in potential acidity (H+Al) compared to the control; compared to the initial values, in all the experimental units potential acidity decreased. However, with the exception of the control unit, the values for the percentage of aluminum saturation in the experimental units increased when compared to those at the start of planting. As the granite sawdust sludge contains around 15% aluminium oxide (Table 3), during the period of the experiment, part of this material may have been solubilized, thus contributing to the increase in available aluminium in the soil.

Comparing the levels of the cations sodium, potassium, calcium and magnesium in the experimental units at the end of the experiment (Table 7) with those shown in Table 5, it can be seen that, with the exception of calcium, probably due to the presence of this element in the simple superphosphate, the levels of the other cations decreased due to their absorption by the plants.

The application of simple superphosphate in the experimental units contributed to the increase in phosphorus levels in the soil and, as it is a salt, to the increase in EC values (Table 7).

With regard to organic matter, it can also be seen in Table 7 that its values decreased when compared to those in Table 5, probably due to its decomposition over the course of the experiment.

4.2.4. Evaluation of Plant Characteristics

Table 8 shows the average fresh and dry matter yields of the aerial part, dry matter yield of the root, plant height and leaf length.

Table 8. Average production of fresh and dry biomass of stems, sheaths, leaves and roots (average of three replications).

Treat ments	Fresh biomass				Dry biomass		Plant height		Comp. of
	hems	stems	leaves	hems	stem	leaves	roots		
gkg 1·	___1			__-1				"	_"-1
0	15,89	12,34	14,87	3,10	1,78	1,85	16,32	20,60	54,82
0,56	11,49	14,55	13,62	2,50	2,29	2,01	10,19	31,70	53,66
1,2	11,55	12,24	12,11	2,58	1,94	2,01	11,06	26,08	49,48
1,84	10,65	11,31	14,30	3,48	1,72	1,98	11,32	21,40	51,65
2,48	15,61	12,11	12,76	3,42	1,87	2,02	12,17	24,75	54,63
3,12	15,74	12,90	15,78	3,38	2,04	1,92	13,82	24,70	47,94
3,76	16,86	12,14	15,80	3,26	2,14	2,31	12,14	26,17	54,73
4,4	14,63	16,30	17,83	3,03	2,41	1,88	11,61	26,40	57,67
5,04	19,74	14,06	19,32	4,02	2,02	2,10	13,06	20,00	59,36
5,68	20,48	14,45	15,78	3,88	2,11	1,96	9,62	22,80	57,93
6,32	24,93	14,99	22,82	4,53	2,16	2,16	13,65	21,25	63,80

Figura 1. Fresh Biomass and Dry Biomass Production

As shown in Figure 9, the total yields (sheath + stem + leaf) of the fresh biomass of the aerial part increased as a function of the treatments, with this effect being quadratic. In addition to the treatments themselves, the soil variables that had a significant effect on these yields at 5% probability were hydrogen in the fresh biomass of the sheath, organic matter and sodium in the fresh biomass of the stem and aluminum in the fresh biomass of the leaf (Table 10).

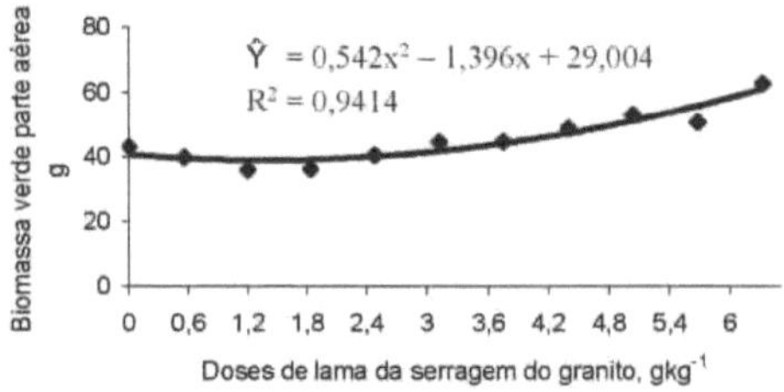

Figura 9. Total fresh biomass production of the aerial part as a function of the treatments.

Similarly, total dry biomass production also increased according to the treatments, as can be seen in Figure 10.

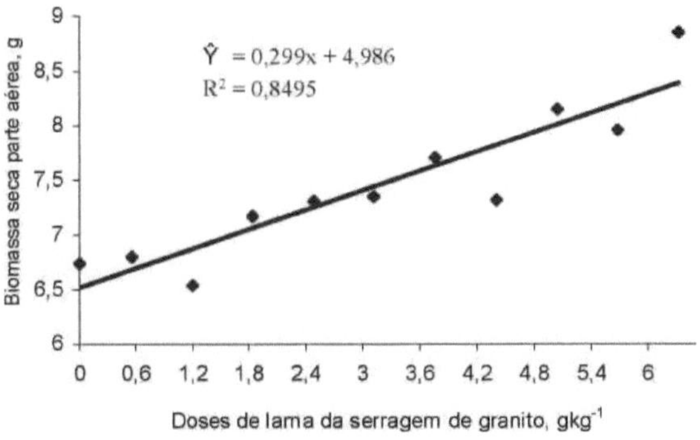

Figura 10. Total dry biomass production of the aerial part as a function of the treatments

Analyzing the results in Table 9, it can be seen that there was a 31.30% increase in dry biomass production in the experimental unit that received the highest dose of slurry compared to the control unit. The accumulation of dry biomass by the maize crop is greatly influenced by the level of soil fertility (Bull and Cantarella, 1993) and according to the results obtained, there was an increase in dry biomass as the soil's CEC increased (Table 7).

The multiple regression analysis (Table 10) shows that dry biomass yields were influenced not only by the treatments, but also by the increased levels of calcium (sheaths), magnesium and aluminium (leaves), sodium and organic matter (stems) present in the soil.

The highest production of dry root biomass was observed in the control (Table 8), contrary to what was expected, since, with the neutralization of aluminium, the tendency was for there to be greater development of the root system in these experimental units. Bull and Cantarella (1993) state that the existence of genetic variability in maize for tolerance to aluminium has been verified in cultivars under field conditions, in soils with high exchangeable aluminium saturation (60%) and high acidity (pH 4.3).

The results in Table 9 show that the aerial part of the control plants was the most affected by the soil characteristics, since this experimental unit had the lowest aerial/root ratio. Apparently, the potential acidity of the soil did not have a direct effect on the shoot-to-root ratio. The highest values were observed in the experimental units that received the highest doses of sludge, which may have been due to the greater availability of nutrients for the plants in these experimental units.

Table 10 shows that the soil factors that most influenced the development of roots and leaf area were granite sawdust sludge and the potassium and aluminium content of the soil.

Table 9. Production of dry biomass of the aerial part and roots, aerial part/root ratio, leaf area ratio and leaf area of the plants (average of three repetitions).

Treatments	Area leaf	Total dry biomass	Root dry biomass	Aerial part/root ratio	Leaf area ratio
	cm^2 pot 1^-	$^{-1}$ gvaso^{-1}		cm g^{2-1}	cm g^{2-1}
0	132,99	6,74	16,32	8,14	19,73
0,56	120,32	6,80	10,19	11,81	17,69
1,2	113,49	6,54	11,06	10,26	17,35
1,84	128,13	7,17	11,32	11,32	17,87
2,48	126,78	7,31	12,17	10,42	17,34
3,12	136,15	7,35	13,82	9,85	18,52
3,76	138,18	7,70	12,14	11,38	27,95
4,4	160,03	7,32	11,61	13,78	21,86
5,04	163,07	8,15	13,06	12,48	20,00
5,68	159,68	7,96	9,62	16,60	20,06
6,32	174,56	8,85	13,65	12,78	19,72

4.2.4.2 Leaf area

Analyzing the results of the leaf area as a function of the soil pH values, it can be said that, apparently, the leaf area was not affected by the acidity of the soil, since there was an increase in its values even in those treatments with lower pH values (Figure 11). However, apart from the treatments themselves, aluminum was the main parameter that significantly influenced the variation in leaf area at the 5% probability level (Table 10).

Table 10. Multiple regression equations of plant variables as a function of soil variables.

Dependent variable (Y)	Model	R^2
Fresh biomass of sheaths	$\hat{Y}$ = 2.04+0.43Lama+2.28H	0,50
Fresh stem biomass	$\hat{Y}$ = 7.82+3.90M.O. - 65.67 Na	0,28
Fresh leaf biomass	$\hat{Y}$ = 1.72+0.06Lama+0.64Al	0,66
Sheath dry biomass	$\hat{Y}$ = 4.82+0.14Lama +0.21Ca	0,40
Dry biomass of stems	$\hat{Y}$ = 0.78+0.52 M.O. - 9.04Na+0.731Mg	0,36
Dry biomass of leaves	$\hat{Y}$ = 1.65+0.08Lama+0.62Al	0,46
Root dry biomass	$\hat{Y}$ = 0.0216 - 0.388Lama + 13.44K	0,39
Leaf area	$\hat{Y}$ = 104.34+3.61Lama+42.22Al	0,71
Plant height	$\hat{Y}$ = 45.39+0.74Lama+25.45Al	0,63
Leaf length	$\hat{Y}$ = 41.56+0.42Lama+3.37MO.	0,39
Nitrogen in stems	$\hat{Y}$ = 1.34+0.03Lama-2.28C.E.-0.11P	0,58
Phosphorus in stems	$\hat{Y}$ = -0.65+0.04Lama+0.21M.O.+5.56K	0,82
Potassium in stems	$\hat{Y}$ = 0.44+0.04Lama+0.21M.O.+5.26K	0,84

Nitrogen in leaves	$\hat{Y} = 0.38+0.02Lama+3.14K+0.22Al$	0,77
Phosphorus in leaves	$\hat{Y} = 0.11-0.02Lama$	0,31
Potassium in leaves	$\hat{Y} = 4.32+0.02Lama-1.48pH+0.25M.O.+0.49Ca+0.41Mg$	0,92
Nitrogen in roots	$\hat{Y} = 0.54+0.20Lama$	0,36
Phosphorus in roots	$\hat{Y} = 0.18+0.03\ Ca$	0,20
Potassium in roots	$\hat{Y} = 0.92+0.02Lama+3.47K-0.09Mg$	0,75

The leaf area ratio varied by 20.68% considering the highest and lowest values obtained. This variation was irregular and was not directly related to the amount of sludge applied to the soil, but was probably due to the concentration of nutrients in the plants during the experiment.

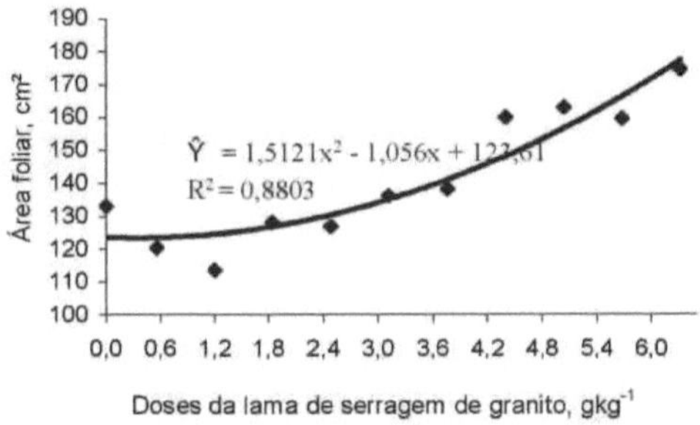

Figura 11. Leaf area production as a function of treatment.

11.2.4.3. Leaf length

Leaf length behaved in the same way as leaf area. Multiple regression analysis showed that its variation was influenced, at a significance level of 5%, by the treatments and the organic matter present in the soil.

11.2.4.4. Plant height

The values observed for plant height show that, although there was an increase compared to the control, the variation found between treatments was small. The main factors influencing this variation were the treatments and the aluminium in the soil.

The low nitrogen content in the biomass analyzed may have been responsible for the variation in plant height. According to Bull and Cantarella (1993), nitrogen deficiency delays cell division at growth points, resulting in a reduction in leaf area and plant size.

11.2.4.5. Plant nutrition

Table 11 shows the nitrogen, phosphorus and potassium content of the stem, leaf and root biomass.

Table 11. Nitrogen, phosphorus and potassium content of stem, leaf and root biomass as a function of granite sawdust sludge doses

Treat	Stem			Leaves			Roots		
gkg^{-1}	N	P	K	N	P - gkg^{-1} -	K	N	P	K
0	5,1	0,9	2,8	6,8	1,0	6,3	5,3	0,6	0,9
0,56	5,8	0,7	2,8	6,6	1,1	4,8	6,9	1,0	1,7
1,2	6,6	1,1	3,5	6,1	1,0	4,8	5,1	0,8	1,5
1,84	5,4	1,1	3,1	6,4	1,0	4,2	5,8	0,8	1,5
2,48	4,4	1,1	3,3	6,4	1,0	5,7	6,0	0,8	1,7
3,12	6,4	1,1	3,5	6,7	1,0	6,8	6,0	0,8	1,5
3,76	6,9	1,0	6,5	7,9	0,8	7,8	6,9	1,0	2,4
4,4	7,7	0,9	6,6	8,8	0,7	1,8	7,6	1,3	3,5
5,04	7,1	0,7	8,3	9,6	0,7	3,1	7,4	0,9	3,5
5,68	8,2	0,9	2,9	10,9	0,9	8,3	7,9	1,0	4,6
6,32	7,5	0,7	9,8	9,0	0,7	5,7	7,6	0,9	3,9

a) Nitrogen

The nitrogen concentrations present in the biomass analyzed (Table 11) showed an upward trend as the amount of sludge incorporated into the soil increased, reflecting the benefits of this incorporation. However, these concentrations are below the minimum concentrations suitable for growing corn (Raij, 1991).

Figure 12 shows that the nitrogen content in the biomass was significantly influenced by the treatments. The coefficients of determination indicate that 60, 91 and 67% of the variation in nitrogen content in the stem, leaf and root, respectively, can be attributed to the variation in doses of granite sawdust sludge.

According to the multiple regression analysis (Table 10), the soil variables that most influenced the nitrogen content of the biomass analyzed, in addition to the treatments, were the potassium and aluminium content (leaves), phosphorus and the increase in electrical conductivity (stems).

As the sludge does not contain nitrogen, the greater accumulation of this nutrient is explained by the favorable effect of the sludge on the general development of the plant, accompanied by greater absorption of nitrogen from the soil itself due to greater availability of the element. This availability is increased by the stimulus that raising the pH gives to bacteria and the nitrification process (Sutton and Monteiro, 1997). Pereira (1999) found similar behavior when using basalt dust on dry matter production and macronutrient absorption by the maize crop.

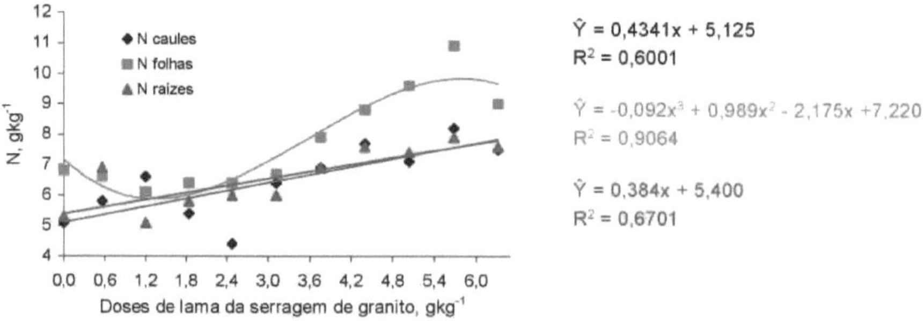

Figura 12. Nitrogen concentration in stems, leaves and roots as a function of treatment.

b) Phosphorus

The phosphorus concentrations in the biomass show that the roots accumulated this nutrient increasingly according to the treatments, but the leaf biomass behaved in the opposite way, i.e. it decreased as the amount of granite sawdust sludge increased. The phosphorus content in the stem biomass increased up to a dose of 3.12 g/vessel, decreasing with the following doses.

The multiple regression analysis (Table 10) indicates that the granite sawdust sludge influenced the phosphorus levels present in the biomass analyzed, along with the organic matter and potassium (stems) and calcium (roots) present in the soil.

As with nitrogen, the concentrations of this nutrient in the plants were below the minimum adequate for the nutritional supply of corn (Raij, 1991), which compromises the growth of the root system and, consequently, the development of the aerial part and grain production.

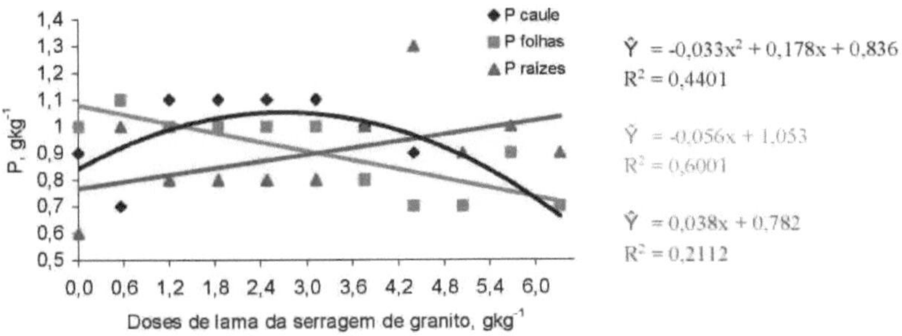

Figura 13. Phosphorus concentration in stems, leaves and roots as a function of treatment.

c) Potassium

The potassium concentrations of the biomass increased as the amount of granite sawdust sludge applied to the soil increased. As granite sawdust sludge has this nutrient in its composition (Table 2), it was released into the soil, as can be seen from the increasing potassium values found in the soil (Tables 5 and 7) according to the increasing doses of sludge. The potassium values in the soil before planting (Table 5) were higher than those obtained after planting (Table 7), which proves that the plants assimilated this nutrient. The multiple regression analysis (Table 10) confirms the influence of the sludge in increasing the potassium levels in the biomass analyzed. However, according to Raij (1991), the minimum potassium content suitable for maize plants (1.16%) was only obtained from the treatment of 3.12 $g.kg^{-1}$.

The multiple regression analysis (Table 10) indicated that, in addition to the sludge doses, other elements influenced the potassium levels in the biomass analyzed, such as organic matter and soil potassium (stems), calcium, magnesium and pH (leaves) and potassium and magnesium (roots).

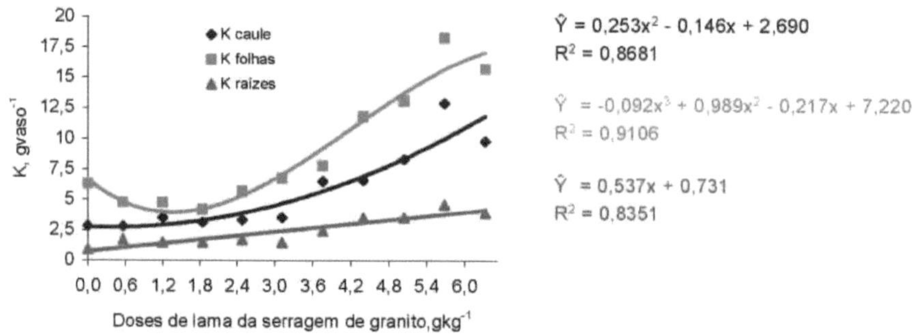

Figura 14. Potassium concentration in stems, leaves and roots as a function of treatment.

Tables 1 and 2 (APPENDIX) show the statistical models obtained.

5-CONCLUSIONS

Considering the conditions of the experiment and the material used, it can be concluded that:

1. Although the pH was not raised to levels suitable for growing most crops, the granite sawdust slurry had a neutralizing effect on soil acidity;

2. After the incubation period, the granite sawdust sludge increased the CEC, phosphorus and organic matter in the soil;

3. The neutralizing effect of the sludge on the soil's acidity decreased over time;

4. The granite sawdust sludge contributed to an increase in potassium levels in the plants;

5. Granite sawdust sludge can be used for agricultural purposes.

6-BIBLIOGRAPHICAL REFERENCES

AGUIAR, A. P. A. **Management of soil fertility under pasture, liming and addition.** Guaiba: Agropecuâria, 1998. 120p.

ALBUQUERQUE, J. A.; BAYER, C.; ERNANI; P. R. and FONTANA, E. C. Physical and electrochemical properties of an Iatossolo bruno affected by liming. Revista Brasileira de Ciência de Solo, Viçosa, 24: 295-300, 2000.

ALCARDE, J. C. **A avaliaçà da qualidade dos corretivos de acidez dos solos.** Campinas: Cargill Foundation, 1986. 40p.

AMARAL, R. D.; BARROS, N. F.; COSTA, L. M. and FONTES, M. P. F. **Effect of zinc industry residues on the chemistry of soil samples and corn plants**. Revista Brasileira de Ciência do Solo, Campinas, v.20, p. 443-440, 1996.

ANJOS, J. T.; PUNDEK, M.; TASSINARI, G. and GRIMM, S. S. **Effect of liming and phosphate fertilization on some chemical properties of a dystrophic Inimic cambisol cultivated with maize**. Revista Brasileira de Ciência de Solo, Campinas, 5: 5054, 1981.

Brazilian Mineral Yearbook. Brasilia, National Department of Mineral Production, v. 27, 1998.

ARAÙJO, A. P. and PERLINGEIRO, B. M. **Characterization of organic residues from salt pans and possibilities for their use as fertilizers.** Revista Brasileira de Ciência do Solo, Campinas, v. 23, p. 443-440, 1999.

AZEVEDO, A. C.; KAMPF, N. and BOHNEN, H. Changes in the evolutionary dynamics of Latossolo Bruno by liming. Brazilian Journal of Soil Science. Campinas, 1: 2831, 1996.

BARBOSA FILHO, M. P.; FAGERIA, N. K.; SILVA, O. F. de and BARBOSA, A. M. **Interactions between liming and zinc on nutrient absorption and rainfed rice production in greenhouse conditions.** Revista Brasileira de Ciência de Solo, Campinas, 16: 355-366, 1992.

BATAGLIA, O. C. **Irrigation system in pots for fertilization experiments.** Revista Brasileira de Ciência do Solo, Campinas, v.14, p. 25-31, 1990.

BRACCINI, M. C. L.; MARTINEZ, H. E. P.; PEREIRA, P. R. G.; SAMPAIO, N. F. and PEREIRA. **Tolerance of coffee genotypes to alluminium in nutrient solution II. P, Ca and Al contents and P and Ca efficiency.** Revista Brasileira de Ciência do Solo, Campinas, v.22, p. 443-450, 1998.

BULL, L. T. and CANTARELLA, H. **Corn cultivation: factors affecting productivity.** Piracicaba: Potafos, 1993. 301p.

CAIRES, E. F.; CHUEIRI, W. A.; MADRUGA, E. F. and FIGUEIREDO, A. **Changes in soil chemical characteristics and soybean response to surface-applied limestone and gypsum in a no-tillage cropping system.** Revista Brasileira de Ciência do Solo, Campinas, v.22, p. 27-34, 1998.

CAIRES, E. F.; BANZATTO, D. A. and FONSECA, A. F. **Surface liming in a no-till system.** Revista

Brasileira de Ciência do Solo, Campinas, v.24, p. 161-169, 2000.

CARVALHO, E. C. **Acidity and liming.** Cuiabà: EMATER-MT, 1989. 60p.

CHAVES, L. H. G. **Principios de quimica e fertilidade do solo.** Campina Grande: UFPB, 1997. 130p. (Handout).

COSTA, A. N.; PREZOTTI, L. C. and LANI, J. A. **Making use of waste from the marble and granite processing industry as a soil improver in Espírito Santo.** Research Report, 1995, 18 p.

CUNHA, G. M.; CATEN, A.; GOMES, J. A.; FREIRE, R. R. and PIRES, F. R. **Chemical characteristics of marble and granite residues and their efficiency as a soil acidity corrector.** In: Reuniao Brasileira de Fertilidade do Solo e Nutriçao de Plantas, 22: 1996, Manaus, AM.

DA SILVA, S. A. C. **Characterization of granite block sawdust. Study of its potential for use in the manufacture of laying mortars and soil-cement bricks.** Vitória: UFES, 1998. 159p. (Master's dissertation).

EMATER. **Fertilization suggestions for the state of Paraiba; 1ª approximation.** Joao Pessoa: EMATER - PB, 1979.

BRAZILIAN AGRICULTURAL RESEARCH Corporation - EMBRAPA. National Soil Survey and Conservation Service. **Soil and limestone analysis methods.** By BLOISE, R. M.; MOREIRA, G. N. C. and DYNIA, J. F. 2ª ed. rev. Rio de Janeiro, EMBRAPA-SNLCS, 1979, 32 p. (Boletim técnico, n. 55).

BRAZILIAN AGRICULTURAL RESEARCH Corporation - EMBRAPA. National Soil Survey and Conservation Service. **Manual and methods for soil conservation analysis.** Rio de Janeiro: EMBRAPA-SNLCS, 1997.

BRAZILIAN AGRICULTURAL RESEARCH COMPANY - EMBRAPA. National Soil Survey and Conservation Service. **Criteria for distinguishing soil classes and mapping unit factors; standards in use by the SNLCS.** Rio de Janeiro: EMBRAPA - SNLCS, 1998.

BRAZILIAN AGRICULTURAL RESEARCH COMPANY - EMBRAPA. National Soil Research Center. **Brazilian Soil Classification System.** Brasilia: EMBRAPA. Information production, Rio de Janeiro. EMBRAPA Solos, 1999.

ERNANI, P. R. and ALMEIDA, J. A. **Evaluation of methods and quantitative recommendation of limestone for the soils of the State of Santa Catarina.** Revista Brasileira de Ciência de Solo, Campinas, 10: 143-150, 1986.

ERNANI, P. R and BARBER, S. A. **Corn growth and changes of soil and root parameters as affected by phosphate fertilizers and liming.** Pesquisa Agropecuària Brasileira. Brasilia, 26: 1309-1314, 1991.

ERNANI, P. R.; FIGUEIREDO, O. R. A.; BECEGATO, V. and ALMEIDA, J. A. **Decrease in soil phosphorus retention by increasing pH.** Revista Brasileira de Ciência de Solo, Campinas, 20: 159-162, 1996.

ERNANI, P. R.; NASCIMENTO, J. A. L. and OLIVEIRA, L. C. **Increase of grain and green matter of corn by liming.** Revista Brasileira de Ciência de Solo, Campinas, 22: 275-280, 1998.

ERNANI, P. R.; NASCIMENTO, J. A. L.; CAMPOS, M. L. and CAMILLO, R. J. **Influence of the combination of phosphorus and limestone on corn yield.** Revista Brasileira de Ciência de Solo, Campinas, 24: 537-544, 2000.

ESCOSTEGUY, P. A. V. and KLAMT, E. **Ground basalt as a source of nutrients.** Revista Brasileira de Ciência de Solo, Viçosa, 22: 11-20, 1998.

FAQUIN, V.; ANDRADE, C. A. B.; FURTINI NETO, A. E.; ANDRADE, A. T. and CURI, N. **Response of the bean plant to the application of limestone in varzea soils of southern Minas Gerais.** Revista Brasileira de Ciência de Solo, Viçosa, 22: 651-660, 1998.

FELLENBERG, G. **Introduction to environmental pollution problems**. Sao Paulo, Editora da Universidade de Sao Paulo, 1980 197 p.

FERNANDES, L. A.; FAQUIN, V.; FURTINI NETO, A. E. and CURI, N. **Phosphorus fractions and acid phosphatase activity in bean plants cultivated in varzea soils.** Revista Brasileira de Ciência de Solo, Viçosa, 24: 561-571, 2000.

FERRI, M. G. **Fisiologia Vegetal.** Sao Paulo, EPU, 1985, 361 p.

FONTES, R. L. F. and RIBEIRO, A. C. **Inorganic residues of industrial origin.** In: Congresso Brasileiro de Ciência de Solo, 27, 1999, Brasilia, Sociedade Brasileira de Ciência do Solo, CD-ROM.

FORESTIERI, E. F. and DE-POLLI, H. **Liming, sulphur and micronutrients on the growth of corn and black mucuna in a red-yellow podzolic.** Revista Brasileira de Ciência de Solo, Campinas, 14: 167-172, 1990.

FRANCHINI, J. C.; MALAVOLTA, E.; MIYAZAWA, M. and PAVAN, M. A. **Chemical changes in acid soils after the application of plant residues.** Revista Brasileira de Ciência de Solo, Viçosa, 23: 533-542, 1999.

GUIMARAES, R. P. **The asymmetry of shared interests: Latin America and the global environmental agenda.** In: LEIS, H. R. **Ecologia e Politica.** Rio de Janeiro, Vozes, 1991.

JOHN, M.; LATOUR, D.; ROUTTEN, E. and BROCKLEBANK, J. **Principles of regression analysis course notes.** U.S.A., 1994, 369 p.

JORGE, J. A.; CAMARGO, O. A. and VALADARES, J. M. A. S. **Physical conditions of a dark red latosol four years after the application of sewage sludge and limestone.** Revista Brasileira de Ciência de Solo, Campinas, 15: 237-240, 1991.

KABEYA, K. S.; PINTO, J. C.; FAQUIN, V.; FURTINI NETO, A. E. and ANDRADE, A. F. **Liming and phosphorus doses on growth and phosphate nutrition of grasses and forages in the establishment phase.** In: Reuniao Brasileira de Fertilidade do Solo e Nutriçao de Plantas, 22, Manaus, 1996. Expanded abstracts, Manaus, Ed. da Universidade do Amazonas, p. 242-247, 1996.

LANTMANN, A. F.; SFREDO, G.J.; BORKERT, C. M. and OLIVEIRA, M. N. **Soybean response to molybdenum at different soil pH levels.** Revista Brasileira de Ciência de Solo, Campinas, 13: 45-49, 1989.

MALAVOLTA, E. **Fertilizers and phosphate fertilization.** Fertilizantes Mitsui S.A. Sao Paulo, 1985, 61 p.

MELLO, F. A. F.; BRASIL SOBRINHO, M. O. C.; ARZOLLA, S.; SILVEIRA, R. J.; COBRA NETTO, A.; KIEHL, J. C. **Fertilidade do solo.** Sao Paulo: Nobel, 1983, 400 p.

MIRANDA, A. S. e MIRANDA, J. C. C. **Efeito residual do calcàrio na produção de milho e soja em glei pouco húmico.** Revista Brasileira de Ciência de Solo, Viçosa, 24: 209-215, 2000.

MIYAZAWA, M.; CHIERICE, G. O. and PAVAN, M. A. **Amelioration of aluminium toxicity to wheat roots by complexation with organic acids.** Revista Brasileira de Ciência de Solo, Campinas, 16: 209-215, 1992.

MORELLI, J. L.; DALBEN, A. E.; ALMEIDA, J. O. C. and DEMATTÊ, J. L. I. **Calcium and gypsum on sugarcane productivity and on the chemical characteristics of a medium-loamy latosol.** Revista Brasileira de Ciência de Solo, Campinas, 16: 187-194, 1992.

NOVELINO, J. O.; LEITE, P. C.; PROENÇA, S. S. and SILVA, M. L. **Chemical characteristics of two latosols from Dourados - MS, submitted to the application of basalt dust**. In: XXV Brazilian Congress of Soil Science, III, 1995, Viçosa. Expanded abstracts, Campinas, Brazilian Society of Soil Science, 1995, p. 1334-1335.

OLIVEIRA, A. J.; GARRIDO, W. E.; ARAÙJO, J. D.; LOURENÇO, S. **Métodos de pesquisa em fertilidade do solo.** Brasilia: EMBRAPA-SEA, 1991. 392p.

PEREIRA, M. C. and KIEHL, J. de C. **Use of basalt crushing residue in agriculture.** In: Congresso Brasileiro de Ciência de Solo, 27, 1999, Brasilia, Sociedade Brasileira de Ciência do Solo, CD-ROM.

PRIMAVESI, A. **Manejo Ecològico do Solo: A agricultura em regiôes tropicais.** Sao Paulo: Nobel, 1990. 549p.

QUAGGIO, J. A.; DECHEN, A. R. and RAIJ, B van. **Effect of applying limestone and gypsum on peanut production and soil base leaching.** Revista Brasileira de Ciência de Solo, Campinas, 6: 189-194, 1982.

QUAGGIO, J. A.; RAIJ, B. van.; MALAVOLTA, E. **Alternative use of the SMP-buffer solution to determine lime requirement of soils. Communications in Soil Science and Plant Analysis.** New York, 1983.

QUAGGIO, J. A. **Soil reaction and its control.** In: DECHEN, A. R. and CARMELLO, Q. A. C. Simpòsio Avançado de Quimica e Fertilidade do Solo, Campinas, Fundaçao Cargill, 1986, p. 53-89.

RAIJ, B. V. **Improving the subsoil root environment.** In: SIMPÓSIO AVANÇADO DE SOLOS E NUTRIÇÂO DE PLANTAS, 2, 1989, Piracicaba. Proceedings. Campinas: Cargill Foundation, 1989. 148p.

RAIJ, B.V. **Fertilidade do solo e adiibaçào.** Piracicaba: Ceres, Potafos, 1991. 343p.

RAIJ, B. van; CAMARGO, A. P.; CANTARELLA, H. and SILVA, N. M. **Exchangeable aluminum and base saturation as criteria for liming recommendations.** Campinas, Bragantia, 1993.

RAIJ, B. van; MASCARENHAS, H A. A.; PEREIRA, J. C. V. N. A.; IGUE, T. and SORDI, G. de. **Effect of limestone and gypsum on soybeans grown on sulphate-saturated acid purple latosol.** Revista Brasileira de Ciência de Solo, Campinas, 18: 305-312, 1994.

RHEINHEIMER, D. S.; PETRY, C.; KAMINSKI, J. and BARTZ, H. R. **Influence of aluminum stress on tobacco plants: I. Effects on the root system, phosphorus and calcium absorption and dry mass accumulation.** Revista Brasileira de Ciência de Solo, Campinas, 18: 63-68, 1994.

ROMAGNOLI. **Granite cutting and polishing process**. Quality stone. Granites and Marbles. Sao Paulo, n. 81, p. 16-20, Apr./May/Jun., 1985.

SCHUNKE, R. M.; RAZUCK, R; KIEHL, J. C. and SANTOS, J. C. C. **Characterization of the by-product of niobium processing and effects on the chemical properties of two soils**. XXIII Brazilian Meeting of Soil Fertility and Plant Nutrition, FertBio 98, Abstracts, 1998. p.664.

SILVA, J. D. **Granites in Brazil. Its role in the industrial, economic and social context in the state of Rio de Janeiro.** Quality rocks. Granites and Marbles. Sao Paulo, n. 91, p. 42-46, Oct./Nov./Dec., 1987.

SUTTON, M. and MONTEIRO, F. A. **Dry matter yield of four alfalfa cultivars in a red-yellow Iatosol subjected to liming and potassium fertilization.** Revista Brasileira de Ciência de Solo, Viçosa, 21: 591-597, 1997.

TEDESCO, M. J. **Analysis of soil, plant and other materials.** 2ª ed. rev. e ampl. Porto Alegre, RS, Department of Soils, UFRS, 1995. 147 p. il. (Technical Bulletin, n. 5).

TEDESCO,M. J., CAMARGO, F. A de O. and GIANELLO, C. **Organic waste of agricultural, urban and industrial origin.** In: Brazilian Congress of Soil Science, 27, 1999, Brasilia, Brazilian Society of Soil Science, CD-ROM.

THOMAS, K. **Man and the animal world.** Sao Paulo, Ed. Companhia das Letras, 1988.

7-ANNEXES

Table 1. Results of the mean squares provided by the analysis of variance with unfolding of the degrees of freedom of the mud factor into polynomial components, estimates of the parameters of the fitted models and their respective coefficients of explanation R .[2]

- Plant characteristics

		Leaf area	Leaf length	Root dry biomass	Dry biomass of leaves	Sheath dry biomass
1st grade component	1	9543,49**	296,0588**	40,578	0,10910**	6,9515**
2nd grade component	1	945,68*	166,6126**	12,275	0,01080	0,5576
3rd grade component	1	645,81	10,2082	3,095	0,00120	0,2092
Regression deviation	7	111,56	20,6301	2,043	0,00280	0,4469
Mud	(10)	1191,59**	61,7290**	1,356	0,01407**	1,0846
Waste	22	164,30	15,3723	1,009	0,00286	0,6009
Estimated Parameters	$\hat{\beta}_0$	123,6047	54,1102	13,4485	0,4100	2,6677
	$\hat{\beta}_1$	-1,0555	-2,5016	-0,9833	0,0286	0,2281
	$\hat{\beta}_2$	1,5121	0,6347	0,1375	------	
	$\hat{\beta}_3$	-------	------		------	
R^2		0,88	0,75	0,03	0,78	0,64

** Statistically significant at 1% probability by the F test.

* Statistically significant at 5% probability by the F test.

Source of Variation	G.L.	Mean Squares			
		Plant dry biomass	Fresh biomass leaves	Fresh biomass sheaths	Fresh plant biomass
1st grade component	1	*12,0204***	*2,3050***	*291,9065***	542,4380**
2nd grade component	1	*0,7834*	*0,3515**	*60,799**	121,7863*
3rd grade component	1	*0,0242*	*0,0374*	*4,0044*	0,1404
Regression deviation	7	*0,1960*	*0,0556*	*14,2216*	6,6070
Mud	(10)	*1,4200*	*0,3112***	*45,6262***	71,0614*
Waste	22	*0,9656*	*0,0468*	*13,5670*	25,8853
Estimated	$\hat{\beta}_0$	*4,9864*	*1,8111*	*14,3040*	29,0041
	$\hat{\beta}_1$	*0,2999*	*-0,0519*	*-0,9324*	-1,3966

Parameters	$\hat{\beta}_2$		*-0,0292*	*0,3834*	0,5426
	$\hat{\beta}_3$		------	------	
R^2		*0,85*	0,85	0,77	0,94

** Statistically significant at 1% probability by the F test.

* Statistically significant at 5% probability by the F test.

Source of Variation	G.L.	Mean Squares				
		Stem nitrogen	Phosphorus stem	Stem potassium	Nitrogen leaves	Phosphorus leaves
1st grade component	1	*0,2514***	0,0010	2,7855**	0,5566**	0,0042**
2nd grade component	1	*0,0087*	0,0044**	0,2645**	0,0577**	0,0001
3rd grade component	1	*0,0137*	0,00094	0,0508	0,1002**	0,0006
Regression deviation	7	*0,0210*	0,00067	0,0602	0,0101	0,0004
Mud	(10)	*0,04206*	0,0010	0,3522**	0,0782**	0,0007*
Waste	22	*0,01190*	0,0005	0,0248	0,0041	0,0003
	$\hat{\beta}_0$	*5,125*	0,836	2,690	7,220	1,053
Estimated	$\hat{\beta}_1$	*0,434*	0,178	-0,146	-2,175	-0,056
Parameters	$\hat{\beta}_2$	------	-0,033	0,253	0,989	
	$\hat{\beta}_3$	------			-0,092	
R^2		0,60	0,44	0,87	0,91	0,60

** Statistically significant at 1% probability by the F test.

* Statistically significant at 5% probability by the F test.

Source of Variation	G.L.	Mean Squares			
		Potassium leaves	Root nitrogen	Root phosphorus	Potassium roots
1st grade component	1	*5,6186***	0,1968**	0,0019*	0,3860**
2nd grade component	1	*0,9015***	0,0064	0,0007	0,01834
3rd grade component	1	*0,2472**	0,0146	0,0001	0,0076
Regression deviation	7	*0,0529*	0,0111	0,0010	0,0076
Mud	(10)	*0,0782***	0,0295*	0,0009	0,0465**
Waste	22	*0,0041*	0,0114	0,0004	0,0064
	^	*6,6643*	5,400	0,782	0,731

Estimated Parameters	$\hat{\beta}_0$				
	$\hat{\beta}_1$	*-4,189*	0,384	0,038	0,537
	$\hat{\beta}_2$	*1,842*			
	$\hat{\beta}_3$	*-0,145*			
R^2		0,95	0,67	0,21	0,83

** Statistically significant at 1% probability by the F test.

* Statistically significant at 5% probability by the F test.

Table 2. Results of the mean squares provided by the analysis of variance with unfolding of the degrees of freedom of the mud factor into polynomial components, estimates of the parameters of the fitted models and their respective coefficients of explanation R .[2]

1- Soil after incubation period

Source of Variation	G.L.	Mean Squares				
		pH	**C.E.**	**M.O.**	**In**	**K**
1st grade component	1	0,027	0,190**	440,768**	0,054**	0,256**
2nd grade component	1	0,884**	0,029**	29,087	0,006**	0,045**
3rd grade component	1	0,142**	0,008*	22,444	0,001	0,004
Regression deviation	7	0,028	0,003	36,737**	0,001*	0,001
Mud	(10)	0,1249**	0,0249*	74,9458**	0,0066**	0,0314**
Waste	22	0,0146	0,0015	9,5072	0,0004	0,0016
Estimated Parameters	$\hat{\beta}_0$	4,6105	0,1644	29,9268	0,0733	0,1638
	$\hat{\beta}_1$	0,5268	-0,0745	1,8160	-0,0031	-0,0223
	$\hat{\beta}_2$	-0,1505	0,0332	------	0,0037	0,0105
	$\hat{\beta}_3$	0,0110	-0,0026	------		-------
R^2		0,82	0,91	0,59	0,90	0,96

** Statistically significant at 1% probability by the F test.

* Statistically significant at 5% probability by the F test.

Source of Variation	G.L.	Mean Squares				
		P	**Ca**	**Mg**	**H**	**Al**
1st grade component	1	*5,490***	1,260**	1,811**	7,463**	0,090**

2nd grade component	1	*0,963*	0,315**	0,018	10,996**	0,263**
3rd grade component	1	*0,095*	0,035**	0,010	0,560	0,109**
Regression deviation	7	*1,104*	0,104**	0,330**	0,452	0,037**
Mud	(10)	*1,4363**	0,2339**	0,4147**	2,2156**	0,0719**
Waste	22	*0,4561*	0,0038	0,0162	0,3705	0,037
	$\hat{\beta}_0$	*1,5298*	1,2831	1,8139	6,1260	0,4619
Estimated	$\hat{\beta}_1$	*0,2027*	0,3956	0,1164	-0,7873	-0,4041
Parameters	$\hat{\beta}_2$		-0,0796	-------	0,1628	0,1166
	$\hat{\beta}_3$		0,0055	-------	-------	-0,0096
R^2		0,38	0,69	0,44	0,83	0,64

** Statistically significant at 1% probability by the F test.

* Statistically significant at 5% probability by the F test.

2- Final solo

Source of Variation	G.L.	Mean Squares				
		pH	C.E.	M.O.	In	K
1st grade component	1	*0,0009*	0,0008	231,941**	0,0052**	0,0125
2nd grade component	1	*0,5752***	0,0002	306,049**	0,0002	0,0036
3rd grade component	1	*0,0810***	0,0001	16,5019	0,0006	0,0193*
Regression deviation	7	*0,0866***	0,0005	28,2625**	0,0004	0,0112*
Mud	(10)	*0,1263***	0,0004	75,233**	0,0009	0,0114*
Waste	22	*0,0075*	0,0005	6,1248	0,0004	0,0108
	$\hat{\beta}_0$	*4,4957*		31,5587	0,0776	0,0592
Estimated	$\hat{\beta}_1$	*0,4262*		-4,0903	0,0063	-0,0641
Parameters	$\hat{\beta}_2$	*-0,1160*		0,8602	-----	0,0355
	$\hat{\beta}_3$	0,0083		-----	-----	-0,0041
R^2		0,52		0,72	0,58	0,17

** Statistically significant at 1% probability by the F test.

* Statistically significant at 5% probability by the F test.

Source of Variation	G.L.	Mean Squares

		P	Ca	Mg	H	Al
1st grade component	1	*19,5748*	1,1849**	2,0688**	4,2401**	0,0239*
2nd grade component	1	*49,2002**	0,0772	0,3086	7,6257**	0,3796**
3rd grade component	1	*0,9514*	0,0251	0,1453	0,2574	0,1245**
Regression deviation	7	*30,0956*	0,1729**	0,0599	0,6350**	0,0340**
Mud	(10)	*28,0396*	0,2497**	0,2942**	1,6568**	0,0766**
Waste	22	*14,3385*	0,0263	0,0679	0,1935	0,0039
Estimated Parameters	$\hat{\beta}_0$	*1,8060*	1,7849	1,5068	5,1824	0,5155
	$\hat{\beta}_1$	*1,3454*	0,0942	0,2961	-0,6755	-0,4383
	$\hat{\beta}_2$	*0,2061*		-0,0273	0,1358	0,1279
	$\hat{\beta}_3$				-------	-0,0103
R^2		0,18	0,47	0,81	0,72	0,69

** Statistically significant at 1% probability by the F test.

* Statistically significant at 5% probability by the F test.

MIX
Papier aus verantwortungsvollen Quellen
Paper from responsible sources
FSC® C105338

Printed by Books on Demand GmbH, Norderstedt / Germany